# Synthesis Lectures on Engineering, Science, and Technology

The focus of this series is general topics, and applications about, and for, engineers and scientists on a wide array of applications, methods and advances. Most titles cover subjects such as professional development, education, and study skills, as well as basic introductory undergraduate material and other topics appropriate for a broader and less technical audience.

Erik Cuevas · Julio Cesar Rosas Caro ·
Avelina Alejo Reyes · Paulina González Ayala ·
Alma Rodriguez

# Optimization in Industrial Engineering

From Classical Methods to Modern Metaheuristics with MATLAB Applications

Erik Cuevas
University of Guadalajara
Guadalajara, Jalisco, Mexico

Avelina Alejo Reyes
Facultad de Ingeniería
Universidad Panamericana
Zapopan, Jalisco, Mexico

Alma Rodriguez
Desarrollo de Software
Centro de Enseñanza Técnica Industrial
Guadalajara, Jalisco, Mexico

Julio Cesar Rosas Caro
Facultad de Ingeniería
Universidad Panamericana
Zapopan, Jalisco, Mexico

Paulina González Ayala
Facultad de Ingeniería
Universidad Panamericana
Zapopan, Jalisco, Mexico

ISSN 2690-0300　　　　　ISSN 2690-0327　(electronic)
Synthesis Lectures on Engineering, Science, and Technology
ISBN 978-3-031-74026-8　　ISBN 978-3-031-74027-5　(eBook)
https://doi.org/10.1007/978-3-031-74027-5

© The Editor(s) (if applicable) and The Author(s), under exclusive license to Springer Nature Switzerland AG 2025

This work is subject to copyright. All rights are solely and exclusively licensed by the Publisher, whether the whole or part of the material is concerned, specifically the rights of translation, reprinting, reuse of illustrations, recitation, broadcasting, reproduction on microfilms or in any other physical way, and transmission or information storage and retrieval, electronic adaptation, computer software, or by similar or dissimilar methodology now known or hereafter developed.
The use of general descriptive names, registered names, trademarks, service marks, etc. in this publication does not imply, even in the absence of a specific statement, that such names are exempt from the relevant protective laws and regulations and therefore free for general use.
The publisher, the authors and the editors are safe to assume that the advice and information in this book are believed to be true and accurate at the date of publication. Neither the publisher nor the authors or the editors give a warranty, expressed or implied, with respect to the material contained herein or for any errors or omissions that may have been made. The publisher remains neutral with regard to jurisdictional claims in published maps and institutional affiliations.

This Springer imprint is published by the registered company Springer Nature Switzerland AG
The registered company address is: Gewerbestrasse 11, 6330 Cham, Switzerland

If disposing of this product, please recycle the paper.

# Preface

In an era where efficiency and optimization have become the mantra of engineering and management, it is vital to have the right optimization tools and techniques at our disposal. The rising surge of technology and data analysis has highlighted the unquestionable value of optimization algorithms. These algorithms allow us to find the best possible solutions in a variety of contexts, from resource planning in engineering to strategic decision-making in management.

Learning numerical optimization can be a challenge, as it requires knowledge of mathematics and computer programming. Moreover, the amount of information that has been recently generated and the use of specialized terms can be overwhelming. Especially for those who do not have an expert teacher to guide them in this subject.

This book tackles the following challenge: "To introduce the reader, with light and enjoyable reading, to the world of numerical optimization, while simultaneously providing mathematical and programming tools, in a sequential manner that does not require specialized prior knowledge or the aid of an expert instructor."

The light reading sets this book apart from other specialized books on numerical optimization, where the density of theoretical, mathematical, and programming content makes understanding a few pages take a good amount of time. This will make a considerable difference for readers who are first introduced to the subject and do not have an expert to guide them. Equations and mathematical topics are introduced gradually for better understanding. The main challenge is that the reader does not need to take separate courses or read previous books on mathematics, optimization, and programming to delve into the world of optimization, and once there, from the starting point provided by this book, can decide on routes of greater specialization.

Both the problems and the mathematical expressions and programming codes in Matlab start as simply as possible and increase in level as the chapters progress, to end with an elegant style, mathematically correct and with sophisticated programming. With explanations of each of the steps for correct comprehension.

If you are a student approaching numerical optimization for the first time, you will find this book greatly helpful for your studies, providing you with a solid understanding of optimization algorithms and how they are used in practice. If you are a non-specialized professional with a need to delve into this field, this book will serve as an excellent starting point.

The path to efficiency and optimization can be complex, but with the right tools and techniques, it becomes much more manageable. We hope this book will be a useful companion on that journey.

| | |
|---|---:|
| Zapopan, Mexico | Avelina Alejo Reyes |
| Guadalajara, Mexico | Erik Cuevas |
| Zapopan, Mexico | Paulina González Ayala |
| Zapopan, Mexico | Julio Cesar Rosas Caro |
| Guadalajara, Mexico | Alma Rodriguez |

# Contents

1 **Introduction and John's Story** .................................... 1
   1.1   What and Where Do We Optimize .......................... 1
   1.2   Johns Problem, Introduction to the EOQ ..................... 3
   1.3   Introduction to Solution Methods "the Gradient Descent" ......... 7
   1.4   Solving Jonh's Problem Using the Gradient Descent Method
        with Matlab .......................................... 9
   1.5   Explanation of Code 1.1 ................................. 12
   1.6   Homework Example—Maximizing an Area with a Fixed
        Perimeter ............................................ 15
   References ................................................ 18

2 **Basic Concepts of Optimization** ................................. 21
   2.1   The Zero-Derivative Method ............................. 21
   2.2   Calculating the Gradient When It's not Defined ............... 22
   2.3   The Stopping Criterion .................................. 24
   2.4   Grapich Functions: Plotting the Function and the Current Solution ... 25
   2.5   Plotting Partial Costs and the Total Cost ..................... 28
   2.6   When the Gradient Descent Method Does not Work ............ 30
   2.7   The Random Search Algorithm ........................... 35
   2.8   Modality and Dimensionality ............................. 39
   2.9   Homework Example—Maximizing the Force Among Charges
        with the Gradient Descent Method ......................... 41
   References ................................................ 44

## 3 The Gradient Descent Method Generalization for N Dimensions ... 45
3.1 Introduction ... 45
3.2 The Gradient Descent Method ... 45
3.3 Finding the Peak of the 3D Hill ... 49
3.4 Maximizing the Peaks Function ... 55
3.5 Homework Example—Programming the Gradient Descent Method to Minimize the Bohachevsky Function ... 62
References ... 64

## 4 Curve Fitting ... 67
4.1 Introduction to the Curve Fitting ... 67
4.2 The Squared Error ... 74
4.3 What if the Line Does not Start at the Origin ... 78
4.4 What if the Equation is Quadratic (or a Larger-Order Polynomial) ... 81
References ... 85

## 5 Brief History and Classification of Metaheuristic Optimization Methods ... 87
5.1 Introduction ... 87
5.2 Optimization Methods Classification ... 89
5.3 Exploration and Exploitation ... 91
5.4 Basic Selection Techniques ... 92
References ... 97

## 6 The EOQ Problem with Multiple Suppliers, Restrictions, and Volume Discounts ... 99
6.1 Introduction to the Inventory Administration Problem ... 99
6.2 Example 1. Two Suppliers Without Capacity Constraint ... 101
6.3 Supplier Combination ... 109
6.4 Introducing Capacity Constrains ... 116
6.5 Introducing Volume Discounts ... 119
6.6 Introducing Quality Constrains ... 122
References ... 124

## 7 Probability Distributions and the Random Search Method ... 125
7.1 Introduction ... 125
7.2 Probability Distributions ... 126
7.3 The Random Search Algorithm ... 133
7.4 The Adaptive Random Search Method ... 139
7.5 Homework Example: Maximizing the Area with Fixed Perimeter ... 141
7.6 Homework Example: Maximizing the Peaks Function ... 143
7.7 Homework Example: Minimizing the Bohachevsky Function ... 148
References ... 151

# Contents

**8 The Simulated Annealing Method** .................................... 153
    8.1 Introduction .................................................. 153
    8.2 Description of the Simulated Annealing Method .................. 154
    8.3 Example of the Simulated Annealing Method ...................... 155
    References ....................................................... 157

**9 The Particle Swarm Optimization Method** ........................... 159
    9.1 Introduction .................................................. 159
    9.2 Description of the PSO Method ................................. 160
    9.3 Useful Functions in Matlab .................................... 164
    9.4 Maximizing the Peaks Function with the PSO Algorithm .......... 166
    9.5 Minimizing the Bohachevsky Function ........................... 170
    References ....................................................... 172

**10 Evolutionary Strategies (ES)** ..................................... 173
    10.1 Introduction ................................................. 173
    10.2 The $(1 + 1)$ ES ............................................. 174
        10.2.1 Initialization ....................................... 174
        10.2.2 Mutation ............................................. 174
        10.2.3 Selection ............................................ 176
    10.3 Simulation of the $(1 + 1)$ ES ............................... 176
        10.3.1 Methodology of the $(1 + 1)$ ES Algorithm ............ 177
    10.4 Implementation of the $(1 + 1)$ ES Algorithm in MATLAB ....... 179
    10.5 Variants of Evolutionary Strategies .......................... 181
        10.5.1 Adaptive $(1 + 1)$ ES ................................ 181
        10.5.2 The $(\mu + 1)$ ES ................................... 188
        10.5.3 The $(\mu + \lambda)$ ES ............................. 198
        10.5.4 The $(\mu, \lambda)$ ES .............................. 202
        10.5.5 The $(\mu, \alpha, \lambda, \beta)$ ES ............... 206
        10.5.6 Adaptive $(\mu + \lambda)$ ES and $(\mu, \lambda)$ ES . 207
    References ....................................................... 217

# Introduction and John's Story

Optimization is a natural act for humans, so natural that we do it all the time, sometimes without even realizing it. We can define optimization as follows:

> Optimizing means making decisions to achieve the best possible outcome, considering the resources and constraints we have.

## 1.1 What and Where Do We Optimize

People acquire homes, computers, cars, clothing, and food; they select a profession, a partner, and a city to live in. In all these matters, we prefer the best options over the worst. However, we are usually aware of our limitations and constraints, and because of this, we do not always choose the best options in everything, but rather those that we consider the best decision, taking into account our priorities, our resources, and our constraints. We all agree that a large, beautiful, well-located house is the best option, but sometimes, due to our constraints, we sacrifice size for a good location or vice versa. We do not choose the best options; we choose the *optimal* options, which, from a certain point of view, are the best, considering the options, priorities, and constraints. One way to encompass this task is to call it *Optimization*.

The limitations and constraints best understood are those related to economics, which we understand well because we experience them all the time. That is why this book will introduce the most important concepts by addressing examples from simple economics (without delving into complex financial topics).

Another way to define it would be to say that optimizing, in a sense, can be translated as deciding what cost we are willing to pay or what sacrifices we are willing to make in order to obtain the corresponding benefits. In physical situations or mathematics, something similar happens: there is usually a cost for each outcome, a variable that can be explained as the cost of another. The cost is not necessarily bad; it simply is what we need to do to obtain the expected result.

For example, in the automotive market, car designers offer options for sports cars, vehicles with considerable acceleration capacity. These often come with a high-power engine and are usually small cars with considerably less weight than family cars. We could say that if we want to maximize acceleration, we need to maximize engine power and minimize the car's weight. That's why sports cars are usually small but have very powerful engines. On the other hand, sports car drivers do not expect their cars to have low fuel consumption or maintenance to be economical. Options for cars with low fuel consumption also tend to be small but usually have a low-power engine. Large family cars tend to be considerably big and have a large engine; however, the engine, in relation to the car's weight, is not usually designed to provide high acceleration but to maintain a balance between acceleration and fuel consumption while meeting the constraint of carrying several passengers, sometimes three rows of seats. Cargo trucks usually have a very powerful engine, but due to their size, they tend to be the vehicles with the lowest acceleration in practice.

We can identify clear signals of decision-making in car designers based on guidelines, resources, and constraints.

Would it be possible to design a cargo truck or trailer with high acceleration? Yes, it's possible to manufacture a very large engine. Probably, they do not do it because manufacturers calculate that the cost would be very high, it would not have a high sales volume, and their buyers would prefer the type of trucks that are currently sold. We could review many examples and notice optimization behind almost any human activity.

Optimization isn't only present in car manufacturers but also in individuals or consumers. There are people who prefer not to invest too many resources in a car in order to have those resources available or to allocate them to other expenses they consider more important, or they enjoy more. In these expenses, these individuals try to optimize their resources. Optimization is a widely studied topic [1, 2].

Physical situations can be represented in mathematical terms, what we commonly call "modeling" [3] and once they are modeled mathematically, it is possible to optimize them using the techniques we will describe in this book. In mathematical terms, optimizing implies finding the minimum or maximum values of an objective function under certain mathematical constraints (if any). The objective function is a variable that depends on other input variables. The goal of optimization algorithms is to find not only the minimum or maximum of the objective function but also the values of the independent variables that result in said minimum or maximum value [4].

Revisiting the definition we announced about optimization: "making decisions to achieve the best possible result with the resources and constraints we have," achieving the best results can refer to many things, for example, maximizing profits in a project or minimizing the cost of a product.

As has been mentioned, optimization is used in a wide spectrum of applications, not only those related to economics, but the options related to economics have an outstanding educational appeal so that they will be used for various explanations in this book.

The objective function is a mathematical representation of the variable (for example, cost) that one wishes to optimize [5]. In an optimization problem, the goal is to find the minimum or maximum value of this function. The constraints, on the other hand, are the limitations or conditions under which the problem must be solved.

Optimization problems can be linear or nonlinear [6], depending on whether the objective function and the constraints are represented by a linear equation or not. The objective function can have a convex or non-convex shape. They can also be single-variable or multi-variable, depending on the number of decision variables involved, they can be unimodal or multimodal. These terms, which may sound strange at the moment, will be addressed in the following chapters, but we will start with simple examples that help us gradually wrap our heads around these topics before gradually increasing the difficulty level.

## 1.2 Johns Problem, Introduction to the EOQ

We will start by discussing what we will call John's problem. John lives in a remote community and is the local distributor of purified water, which he sells in one-gallon containers. John buys the gallons for 1 USD each, but to purchase them, he must travel to the nearest town, which costs him 10 USD considering fuel and the maintenance of his car, which has a maximum capacity of 200 gallons.

John sells 5 gallons of water per day, storing them in the local community's warehouse, where he must pay 5 cents for storage per gallon per day.

Routinely, when his inventory runs out (the gallons available for sale), he goes to town and buys the 200 gallons of water that fit in his car to replenish his stock, which lasts 40 days. On that day, John spends 210 USD, 200 for the gallons and 10 for the trip.

John is a curious person who enjoys mental challenges, and he likes to make wise decisions. Sometimes, John wonders if his method of work is the best or if there might be a better way to manage his resources.

So, we will take John's problem and try to optimize it. Specifically, we will try to minimize the cost of maintaining his inventory. Note that we do not mention the price at which John sells his gallons of water; in fact, it is not necessary to know this to minimize the cost of inventory management. However, if the reader wishes, they can imagine a certain selling price, for example, 4 USD per gallon. Also, note that we have defined that he will sell 5 gallons daily, assuming the demand is constant, which we can do as a first approximation. In real life, it is difficult to have an example with constant demand, but we can obtain an average of daily sales and use that parameter for the mathematical model. This is a common practice in various industrial sectors. Moreover, there are algorithms that can help us predict variables like this, but the suggestion is that we focus on John's expenses, which is what we want to minimize with the data that are provided.

The first step to solving this problem is to model it mathematically, that is, to mathematically express the average cost that John pays to maintain his inventory.

Every time he goes to town, John buys $Q$ gallons of water (at this point, $Q = 200$). He knows he must buy at least 5 gallons of water; otherwise, his trip would not even serve to maintain his inventory for a day. He also knows it does not pay to buy too few since going to town costs money. He might wish, for simplicity, to buy in multiples of 5 to know how many days his inventory will last using natural numbers, although this is not mathematically necessary. What we can affirm is that the period $T_C$ with which he must make the trip to buy $Q$ gallons can be expressed as (1.1).

$$T_C = \frac{Q}{d}. \tag{1.1}$$

where $d$ is the demand for gallons per day, which in the studied example is $d = 5$ gallons per day (the amount of gallons he sells daily in his town). Thus, if he purchases $Q = 5$ gallons, he will need to make a trip every $T_C = 1$ day. If he purchases $Q = 15$ gallons, he will need to make a trip every $T_C = 3$ days, and so on. Currently, he purchases $Q = 200$ gallons, which means he goes to town every 40 days.

$T_C$ is called the order cycle period. All companies buy their products and plan their purchases in a cyclical manner; the period may last 15 days or 1 month, and there will be companies whose order cycle for some products lasts several months.

After knowing how often he goes to town (every $T_C$), we can assert that the average cost of transporting his gallons of water from the town to his community can be expressed as (1.2).

$$\text{Average Transportation Cost} = \frac{k}{T_C} = \frac{d}{Q}k. \tag{1.2}$$

where $k$ is the cost of the trip to town ($k = 10$ in the current example). The average transportation cost is $k = 10$ USD per trip divided by 40 days (since currently $T_C = 40$). This amounts to an average of 25 cents per day. If he made trips to bring 100 gallons of water instead of 200, he would have to go to town every 20 days instead of every 40 days, and the average cost would be 0.50 USD per day. Thus, the transportation cost increases when the quantity ($Q$) of gallons he purchases decreases.

Now, let's analyze the storage cost of the gallons of water in the warehouse of his locality. Figure 1.1 shows the inventory behavior; suppose that $Q = 25$, and each container with the symbol "5x" represents 5 gallons. Then, on the first day, John must pay $Q$ times

## 1.2 Johns Problem, Introduction to the EOQ

**Fig. 1.1** Behavior of the inventory

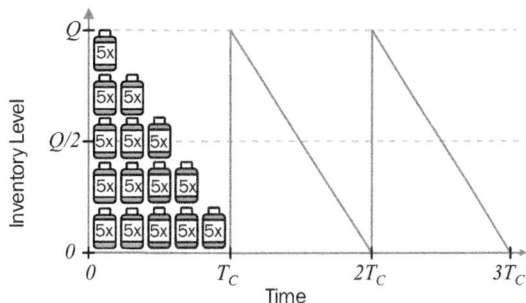

5 cents per each of the 25 gallons. On the second day, he only needs to pay for 20 gallons, and so on, until the gallons are depleted. Every day, John needs to pay less money for storage. Indeed, the area under the curve of the inventory level, multiplied by the inventory cost, is the storage cost. Over an order cycle period, this area can be expressed as (1.3).

$$\text{Area under the curve} = \frac{T_C Q}{2}. \tag{1.3}$$

If we multiply it by $h$ (the storage cost per gallon) and divide it by $T_C$, we will obtain the average cost. This average cost of storing items in the inventory can be calculated as (1.4).

$$\text{Average Storage Cost} = \frac{Q}{2}h. \tag{1.4}$$

From (1.4) it is possible to observe that the average storage cost increases when the quantity ($Q$) of gallons he purchases increases.

The average cost $P$ of the items is fixed (1 dollar per gallon). If we know the daily demand, we can calculate how much John spends on average per day on water gallons, and we can calculate it as (1.5).

$$\text{Average cost of the gallons} = Pd. \tag{1.5}$$

The cost of the gallons does not depend on the quantity ($Q$) of gallons he purchases each time he goes to town. It only depends on the demand, which, in our example, is constant.

The equation for John's average expense can be obtained by adding the three types of costs described by (1.2), (1.4), and (1.5), and can be expressed as (1.6).

$$f(Q) = \frac{d}{Q}k + \frac{Q}{2}h + dP. \tag{1.6}$$

This is the most important formula for our problem, the one that gives us the result we want to minimize. We will call it the objective function $f(Q)$, and since we want to

**Table 1.1** Parameters used to evaluate (1.6)

| Demand | $d = 5$ gallons per day |
|---|---|
| Transportation cost | $k = 10$ USD per trip to town |
| Storage cost | $h = 0.05$ USD per gallon per day |
| Cost per gallon | $P = 1$ USD per gallon |

make it as small as possible, we are facing a minimization problem. Table 1.1 shows a summary of the parameters in (1.6).

Currently, since John makes trips to buy 200 gallons, we can evaluate the objective function (1.6) when $Q = 200$ and with the parameters from Table 1.1. This would be (1.7).

$$f(200) = \frac{5}{200}10 + \frac{200}{2}0.05 + 5 \times 1 = 10.25. \tag{1.7}$$

The result is that $f(200) = 10.25$, meaning John spends 10.25 dollars daily to maintain his inventory. This already considers the cost of transportation, storage, and the price of the water gallons. John pays this because he decides to travel each time his inventory runs out, buying all the gallons that fit in his car, thus going every 40 days for 200 gallons.

If one day he changed his mind about his plan and decided to travel every 20 days to buy 100 gallons, the average inventory cost could be evaluated with Eq. (1.6) with $Q = 100$ and the parameters from Table 1.1, as (1.8).

$$f(100) = \frac{5}{100}10 + \frac{100}{2}0.05 + 5 \times 1 = 8. \tag{1.8}$$

The cost $f(100) = 8$. In other words, John would spend less money, 8 instead of 10.25 dollars daily. An observation we could make is that the transportation cost increased by (1.8) compared to (1.7); however, the storage cost decreased. The decrease in storage cost was greater than the increase in transportation cost; hence, the overall function yielded a smaller result.

Jumping ahead a bit, the optimum decision would be to buy 45 gallons per trip. With this, his inventory cost would be $f(45) = 7.2361$ dollars daily. This decision is the global optimum for this problem, meaning there is no other value of $Q$ that gives a lower cost; the best decision is $Q = 45$.

It's interesting to see how a seemingly insignificant decision can improve the outcome. It would be more interesting to know how to do this.

The difference between the costs described might seem small to some readers, and in this example, it is a small amount, but there are companies whose inventory cost is measured in millions of dollars, and saving a small percentage could mean a lot of money. Moreover, in a competitive world, any advantage counts.

The most important thing is that all John has to do to save money is make a decision and decide how many gallons he should buy each time he goes to town, but how could

John know this? How to find the optimum value? Can it be done with Matlab? Can everything be optimized? Or what things can we optimize? These are the questions we will address in this book.

Optimization plays a central role in many fields, but it is especially relevant in engineering and management. In engineering, optimization problems frequently arise in areas such as planning and process control, system design, resource distribution, and strategic decision-making. Engineers often need to find the most efficient or cost-effective solution to a problem, given certain constraints. This might involve, for example, minimizing the production cost of a certain product while maintaining a specific quality level.

Optimization problems are also very common in management. For instance, a company may want to maximize its profits subject to budget constraints, or it might want to find the most efficient way to allocate its employees to different tasks. Optimization can also be useful in strategic planning, helping managers make informed and efficient decisions.

In the subsequent chapters, we will explore in detail different types of optimization problems and how they can be solved using specific algorithms and techniques. We will also see how these concepts can be applied in engineering and management and how they can be implemented in the Matlab programming language.

Matlab is a powerful numerical computing platform that is also a commonly used program for engineering students. By the end of this book, the reader will have a clear starting point on the fundamentals of optimization and be able to apply these concepts in their own work.

Moreover, with the concepts described in this book, the reader could implement them in other computational packages, such as Excel or Python.

## 1.3 Introduction to Solution Methods "the Gradient Descent"

There are various numerical optimization methods [7, 8], some of which will be addressed in this book. However, we can start with one of the most used and simplest, the *gradient method*, also known as the *gradient descent method*.

To introduce the meaning of these methods, we can imagine that we are on a huge hill that we climb by walking. However, come the afternoon, the top of the hill is covered in dense fog, and we cannot see into the distance, only a few meters ahead of us. We want to descend the hill, at least enough to avoid the fog, so we must define a strategy; we can follow this simple rule: *always go downwards*. We can take a step in each direction, feel which of those steps takes us downhill more quickly, and follow that direction. This, in essence, is the gradient descent method (Fig. 1.2).

The described example deals with a minimization problem. We want to get down or minimize the height at which we are, even though we do not know the way; intuitively, we know that if we always move in the direction of descent, we will surely reach the base of the hill. The counterpart to minimization would be reaching the summit, which

**Fig. 1.2** How to descend a hill if it has been covered in dense fog

would be a maximization problem, and the analogous strategy would be *always to move upwards*.

Returning to our minimization problem, the method requires having a starting point and calculating the "gradient" or the slope at that point, which is a direction in which the function increases most rapidly. Then, we take a step in the opposite direction (if we are trying to minimize the function) or in the same direction (if we are trying to maximize the function). We repeat this process several times until we reach the lowest point, which, in the example of the hill, would be reaching the base.

Each step we take is called an iteration. In general, the methods we will study are iterative [9–11], that is, step by step. In each iteration or step, the same procedure as the previous one is performed, but the values of the variables are updated. The size of the steps we take is important. If we take very small steps, for example, ten centimeters, the method can be very slow. When trying to minimize a function on a computer, our steps can be any value, for example, two kilometers, but if we take very large steps, we might end up on another hill instead of reaching the base of the hill where we are. This step size is often adjusted throughout the optimization process to help balance speed and accuracy.

The gradient method [2], also known as gradient descent, is an optimization algorithm that seeks to find the minimum of a function. The idea is to start at a point $x_0$, from which the following points $x_1$, then $x_2$, and so on $x_3$, ..., $x_n$, $x_{n+1}$, etc., are calculated. The initial point can be provided, chosen based on experience, or simply selected at random.

In every iteration, we say that we are at the point $x_n$, and the next point will be called $x_{n+1}$. To apply the method, it is useful to know the objective function $f(x)$ (for example, Eq. (1.6) for John's problem), but it is not imperative. What is necessary is to know its gradient $\nabla f(x)$ (the gradient symbol is called nabla). The gradient can also be referred to as the slope or derivative in this case.

The term gradient can be used for multi-dimensional functions so that it can encompass a broader meaning. However, in our application, it is simply the slope ($\nabla f(x)$) or the derivative of a function $f(x)$ concerning a variation in $x$. The method involves using this slope $\nabla f(x)$ to calculate the position after the next step, and it is possible to use a parameter $\alpha$ (alpha) to multiply or scale the step length; a small $\alpha$ will lead us in small steps, while a large $\alpha$ will give large steps.

The basic formula of the gradient method is as follows:

## 1.4 Solving Jonh's Problem Using the Gradient Descent Method ...

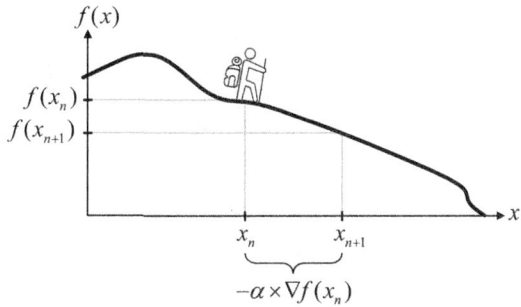

**Fig. 1.3** Variables of formula (1.9)

$$x_{n+1} = x_n - \alpha \times \nabla f(x_n). \tag{1.9}$$

Figure 1.3 shows a drawing with a graphical explanation of Eq. (1.9). The hill is shown in two dimensions, with the x-axis as the horizontal axis and the y-axis as the height $f(x)$.

Suppose we are at iteration (or step number) $n$, our position concerning the x-axis is called $x_n$, and our height is called $f(x_n)$, the height is the objective function in this problem, and the goal is to minimize it.

To achieve our goal, we have to move in the positive direction of the x-axis. The formula contains a negative sign, but the gradient (or derivative) of the objective function in this case is negative. The formula will increase the variable $x$, bringing us closer to the goal. In other words, the term $\alpha \nabla f(x_n)$ is negative, so the negative sign in the formula ensures that the next step of the solution will advance in the direction that naturally minimizes $f(x)$. In a maximization problem, the sign of (1.9) would need to be changed to positive.

Note that the gradient or derivative $\nabla f(x_n)$ can be difficult to obtain for certain applications, so it is sometimes obtained numerically instead of determining it mathematically. Sometimes, the objective function is not derivable, which is why other methods have to be used, which will also be covered later.

## 1.4 Solving Jonh's Problem Using the Gradient Descent Method with Matlab

Going back to Jonhs problem, we already have a $f(x)$ (1.6), well it was an $f(Q)$, but we can easily transform it to an $f(x)$ as (1.10). Let us use code written in Matlab to minimize the average inventory maintenance cost (John's problem), as described in the following expressions.

$$\text{(P)} \quad \text{Min} \quad f(x) = \frac{50}{x} + \frac{0.05x}{2} + 5. \tag{1.10}$$

$$\text{s.t.} \quad 0 < x \leq 200. \tag{1.11}$$

Note that (1.10) is the same function expressed in (1.6), but substituting $Q$ with an $x$ and also substituting the parameter values ($d = 5$, h $=0.05$, k $= 10$).

This is a standard way to describe an optimization problem. It specifies that it is a minimization problem with the word Min, specifies the objective function (1.10), and below this, the constraints are specified (if any). In this case, we specify that the order of gallons ($x$) must be positive (greater than zero); we cannot buy a negative quantity of gallons, and that the maximum value is 200, as it was mentioned that John's car has a maximum capacity of 200 gallons.

An important observation is that in the following problems, we will not specify that $x$ must be a natural number (an integer). However, we can solve it considering the continuity of (1.10), and in the end, consider that the solution will be the integer closest to the result obtained by the code.

The following Matlab code solves John's problem using the gradient descent method. The code is presented and will be explained in its simplest form, to then modify and improve it.

```
% Code 1.1 - Problem 1.0, gradient descent
clear; clc; close;   % reset

f = @(x) 50./x + 0.05*x./2 + 5 ;     % Function
grad_f = @(x) -50/x^2 + 0.025 ;      % Gradient

% Optimizer parameters
max_iter = 200 ;     % maximum number of iterations
q = 20 ;             % starting point
alpha = 100 ;        % learning rate

for i = 1:max_iter   % iterations
    q = q - alpha * grad_f(q) ;
end

Solution = q
```

To run Code 1.1, it's necessary to open Matlab and open a new.m file or script, which can be done by clicking as shown in Fig. 1.4 or pressing Ctrl + N.

An untitled new file will appear. In the editor, we can copy Code 1.1, and paste it into this window. When saving the file, it's recommended to create a new folder on the hard drive with a name that doesn't contain spaces or dots. Instead of these characters, we can add underscores. For example, the folder name "C:\Course_optimization_1" will not cause any problems, but if we name it "C:\Course optimization 1", the spaces could cause conflicts when opening, closing, or executing the script (accents have been omitted in file names).

1.4 Solving Jonh's Problem Using the Gradient Descent Method …

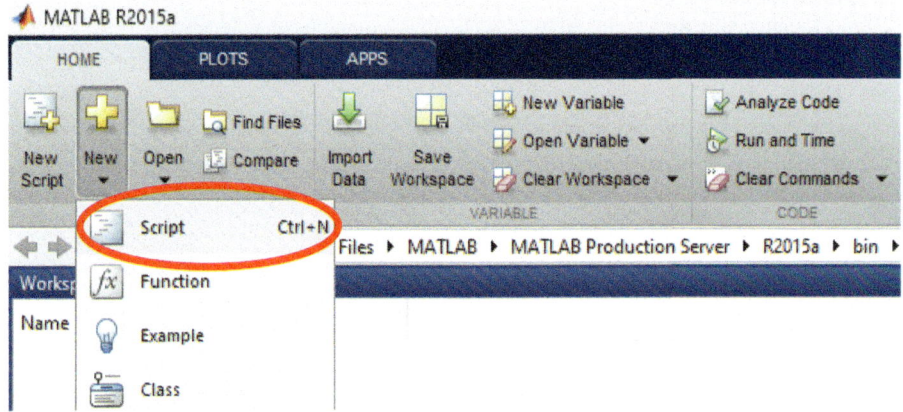

**Fig. 1.4** Opening a new script in Matlab

Moreover, the name of the .m file or script has also the same rules. It can be "code_1_1.m" (the .m extension is added by default), which will not cause any problems, but if we save it as "code_1.1.m", the dot between the ones might cause errors.

If everything goes well, the window should look like Fig. 1.5, and we can run the code with the "Run" button or by pressing F5.

If we create a new folder or use any folder different from Matlab's default folder, a message like the one in Fig. 1.6 may appear. In such a case, it's recommended to click

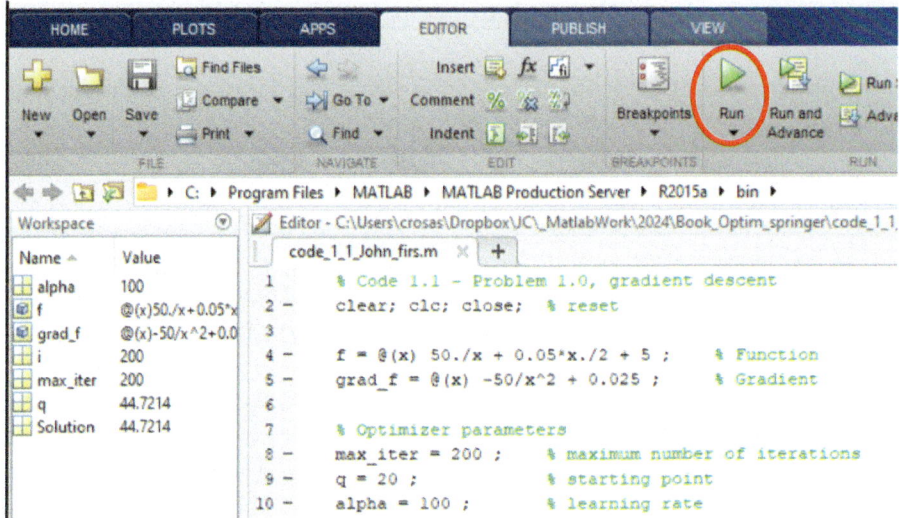

**Fig. 1.5** Running the script file "Code_1_1"

**Fig. 1.6** Message about the folder in which we run the script file

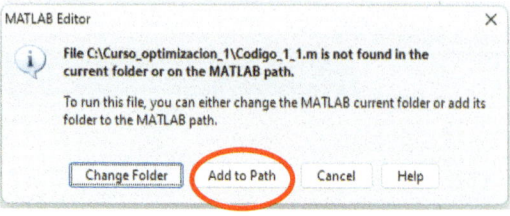

**Fig. 1.7** Command window showing the solution

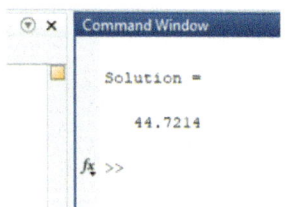

on the "Add to path" option. This message will appear once per session, the first time we run a code or script.

If everything goes well, the .m file or script will display a message in the Matlab command window like the one in Fig. 1.7, telling us that the solution to the problem is to buy 44.72 gallons of water. At this point, we haven't told Matlab that the number of gallons must be an integer, but we can assume the answer is between 44 and 45. As mentioned, the solution to the posed problem is $Q = 45$.

## 1.5 Explanation of Code 1.1

Now, we can address Code 1.1. The first line contains three instructions followed by the word reset.

```
clear; clc; close;    % reset
```

The first instruction (`clear;`) clears the memory; that is, if codes have been executed in Matlab before, Matlab could have stored some variables with assigned values or matrices of variables, so it is advisable to clear the memory to prevent these values from interacting with our program.

The second instruction (`clc;`) clears the command window, although this is not necessary, it helps to keep the workspace tidy. Finally, the instruction (`close;`) closes any graph windows that have been opened beforehand.

After this "reset," we can ensure that our Matlab is clean of previous results and will perform its job according to our code only. The three previously described instructions can be on different lines, for example:

## 1.5 Explanation of Code 1.1

```
clear;
clc;
close;
```

But we can also put them in the same line. The use of the semicolon ";" is not mandatory at this time, but it is a good practice in Matlab to leave a semicolon ";" at the end of code lines to avoid the command window to fill with information we may not need.

Continuing with the code, the next two lines define the objective function and its gradient. The percentage sign (%) indicates a comment that does not execute in the program, but it is useful to put comments in the code that give an idea of what is being done to understand it in a simpler and faster way.

```
f = @(x) 50./x + 0.05*x./2 + 5 ;      % Function
grad_f = @(x) -50/x^2 + 0.025 ;       % Gradient
```

After these definitions, we can call the function `f(x)` and/or `grad_f(x)` and substitute $x$ with some value to evaluate it. For example, in the command window, after executing the program or defining these two functions, we can write "`f(100)`" and press the enter key. The program would respond with $ans = 8$ meaning that the answer to evaluating the function equals 8. The same happens with the gradient function; see Fig. 1.8.

An interesting aspect of Code 1.1 is that it is not really necessary to define the function $f(x)$; it suffices to define its derivative or gradient `grad_f(x)`. In fact, the function $f(x)$ is not used throughout the program. We could say that the code searches for the point whose derivative is closest to zero. However, the function will be used in subsequent codes, and we wanted to include it so that it is clearer to the reader which function we wish to minimize.

Continuing with the code, after defining the objective function and its gradient, the optimizer parameters are defined:

```
% Optimizer parameters
max_iter = 200 ;     % maximum number of iterations
q = 20 ;             % starting point
alpha = 100 ;        % learning rate
```

The `max_iter` parameter defines the maximum number of iterations or steps that the algorithm will perform. The number 200 is relatively low, which is why the code executes in a very short time. Code 1.1 was tested to have a good solution in a short time. It is always important to define a maximum number of iterations; otherwise, a programming error could cause an algorithm to run virtually infinitely, and eventually, the computer's microprocessor could overheat.

We could, for example, change the maximum number of iterations from 200 to 200 thousand, which would make the computer take longer to give us the answer (even if it

**Fig. 1.8** Evaluation of previously defined functions

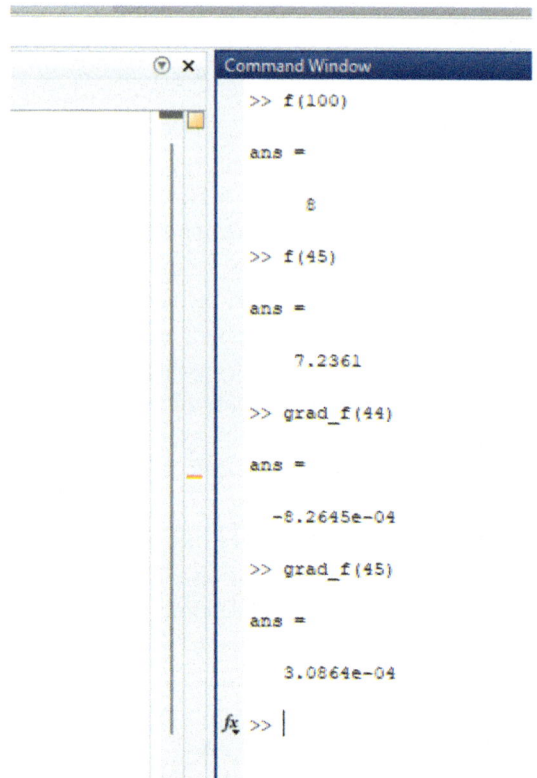

gives the same answer). If we assign many iterations, there is a way to know that Matlab has not finished and is busy doing the iterations. In the lower left part, Matlab indicates that it is executing a process with the word "Busy" (Fig. 1.9).

Additionally, if the algorithm is taking too long and we wish to interrupt it, we can type Ctrl + C, which tells Matlab that it should stop.

On the other hand, if we reduce the number of iterations, for example, from 200 to 20, the code would execute faster; however, it is unlikely to reach the solution, in this case, indicating that the solution is 43.7978 (instead of 44.7214). Despite not reaching the solution, it comes quite close with just a few iterations.

**Fig. 1.9** Indication that a process is running

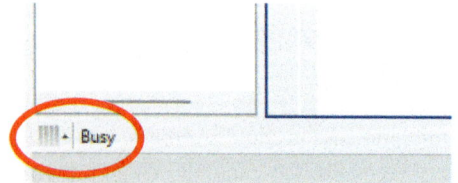

After the number of iterations, we define the initial value of the variable "q = 20;". There is no strict rule about where to start searching for the solution. Often it is done randomly; in this case, it was arbitrarily decided to start with 20 gallons of water.

Finally, the learning rate " alpha = 100;" is defined. This parameter determines the size of the steps in the search process. If it is made too large, as mentioned, it might cause a step to go beyond the space where the solution is located, although if the solution remains within the search space, the algorithm will return to search for the solution. This can also cause delays in finding the solution if the step is very small. 200 iterations might be insufficient. The reader is encouraged to try various values of the learning rate, such as 10, 1, etc.

Lastly, we arrive at the iterations. A "for" loop that is performed the specified number of times, executing Eq. (1.9). In fact, unlike Eq. (1.9), the new variable is updated or overwritten on the previous one; in this case, we do not need to save the previous value, just continue with the iterations.

```
for i = 1:max_iter    % iterations
    q = q - alpha * grad_f(q) ;
end

Solution = q
```

Not adding the semicolon ";" at the end of the instruction is a way to tell Matlab that this instruction should be executed as if it were in the command window and, therefore, display the solution in that window. The last instruction, " Solution = q" is not part of the optimization process, but it allows the solution to be displayed at the end of the screen.

The code helped us to find the solutions. At this point, we can say we can apply the gradient descent method to any problem we can model as an $f(x)$. We will see in the future that we can actually apply this method to cases in which we have two or more input variables, for example, an $f(x_1, x_2)$. But we will also learn the limitations of this algorithm. But no rush; first of all, let us practice.

## 1.6 Homework Example—Maximizing an Area with a Fixed Perimeter

A farmer plans to make a wooden corral for his sheep. He has enough wood to fence a perimeter of 300 m (including the gate) and wants to make it rectangular. With that material, he could make a corral of 50 × 100 m, which would give him a corral of 5000 square meters for his sheep, but he wonders if there's another ratio that would provide more space for the sheep using the same material. To do this, he draws Fig. 1.10 and calls $X$ and $Y$ the dimensions of the corral.

**Fig. 1.10** Area of the land to be fenced

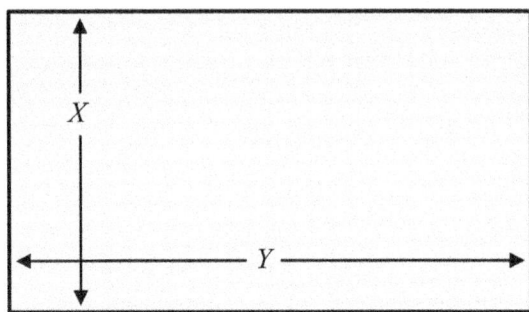

*Answer:*
The goal will be to maximize the area of a rectangle, considering a constraint on the length of its perimeter. The perimeter must not exceed 300 m, and considering the width and height shown in Fig. 1.10, this can be expressed as (1.12).

$$2X + 2Y = 300. \tag{1.12}$$

The farmer can construct a rectangle with a smaller perimeter, so instead of an equality, (1.12) could be an inequality. However, it's logical to think that using a perimeter of less than 300 m would not result in the maximum area value, so (1.12) can be defined as equality. Moreover, the area of the land that is desired to be maximized can be expressed as the product of $X$ by $Y$, see (1.13).

$$A = X \times Y. \tag{1.13}$$

We wish to apply the gradient descent method, which in this case would be gradient ascent since it's a maximization problem. To apply this method, at least the way we have seen it. We also need $f(x)$, which is the objective function, an expression of the area in terms of a single variable. From (1.12), $Y$ can be expressed as (1.14).

$$Y = \frac{300 - 2X}{2}. \tag{1.14}$$

Let's consider the perimeter to be written as $P$ to have the problem written in a general form. The area would be defined in (1.13), which can be rewritten as (1.15).

$$A = X \left( \frac{P - 2X}{2} \right) = \frac{PX - 2X^2}{2} = X\frac{P}{2} - X^2. \tag{1.15}$$

where $P$ is the perimeter, defined as $P = 300$ m for our problem.

Again, we'll use the canonical form to write an optimization problem, which, for the described example, can be expressed as follows:

## 1.6 Homework Example—Maximizing an Area with a Fixed Perimeter

$$\max_{X \in \mathbb{R}} f(X) = X\frac{P}{2} - X^2. \tag{1.16}$$

Subject to:

$$2X + 2Y = 300. \tag{1.17}$$

$$X > 0. \tag{1.18}$$

$$Y > 0. \tag{1.19}$$

The canonical form initially expresses the objective function, with the expression *max* (for maximization) or *min* (for minimization), then writes "subject to" and lists the problem's constraints. And with the constraints, the formulation ends.

The expression below the word max ($X \in \mathbb{R}$) is read as $X$ belongs to ($\epsilon$), the set of real numbers of dimension one ($\mathbb{R}$); in other words, $X$ is a scalar, a number that can take real values, such as *1*, *50*, *60.5*, etc.

There are problems where the solution will not only be a scalar but a set of scalars. For example, if the solution were a pair of coordinates $(x, y)$, then the solution would belong to the set of real numbers of dimension two ($\mathbb{R}^2$).

In the case of the farmer, the solution, or the dimensions of the corral, may be two numbers, X and Y; however, by using equality in the first constraint, it was possible to express Y in terms of X to define the objective function only in terms of X. The constraints (1.18) and (1.19) seem obvious. However, in an optimization formulation, it's important to include all possible information because sometimes the problem modeling can be done by someone expert in a certain topic, and the optimization can be performed by another person specializing in optimization or software or by software, which might not have the physical knowledge of the problem.

Code 1.2 solves the described problem using the gradient method:

```
% Code 1.2 - maximize area, gradient method
clear; clc; close;   % reset

P = 300 ;
f = @(x)   x*P/2 - x^2 ;          % Function
gradient = @(x)  P/2 - 2*x ;

% Optimizer parameters
max_iter = 200 ;      % maximum number of iterations
x_n = 10 ;            % starting point
alpha = 0.1 ;         % learning rate

for i = 1:max_iter  % iterations
    grad  = gradient(x_n) ;
    x_n = x_n + alpha * grad ;
end

Sol_X = x_n
Sol_Y = (P-2*x_n)/2
```

Running the code in Matlab, it is noticeable that the solution is to have $X = 75$ and $Y = 75$. This allows for an area of 5625 square meters, which is 625 square meters more than the initial design the farmer had in mind.

It is exciting to notice that we can use the gradient descent methods for many problems; the Matlab code is relatively easy. It will be used within this book; the reader will also identify the complexity of the problems, and the codes are gradually increasing.

Welcome to the optimization field and to the programming in Matlab. We hope you will enjoy the trip.

## References

1. Rao, S. S., Engineering optimization, theory and practice. Wiley, Fith edition, 2019.
2. Nocedal, J. Wright, S.J. Numerical optimization, Sipringer, 2006.
3. Rsioshansi, Ramteen. Optimization in engineering: Models and Algorithms. SPRINGER, 2019.
4. Deb, K. Evolutionary algorithms for multi-criterion optimization in engineering design. Evolutionary algorithms in engineering and computer science,1999, 2, 135-161.
5. Manchanda, P., Pierre. R., Siddiqi. A.H., Mathematical Modelling, Optimization, Analytic and Numerical Solutions. Springer, 2020.
6. Evtushenko, Y. G., & Stoer, J. Numerical optimization techniques. Optimization Software, Incorporated, 1985, Publications Division.
7. Feng, Z. K., Niu, W. J., & Liu, S. Cooperation search algorithm: A novel metaheuristic evolutionary intelligence algorithm for numerical optimization and engineering optimization problems. Applied Soft Computing, 2021, 98, 106734.
8. Bonnans, J. F., Gilbert, J. C., Lemaréchal, C., & Sagastizábal, C. A. Numerical optimization: theoretical and practical aspects. Springer Science & Business Media. 2006.
9. Tapley, B. D., & Lewallen, J. M. Comparison of several numerical optimization methods. Journal of Optimization Theory and Applications, 1967, 1, 1-32.

10. Dennis Jr, J. E., & Schnabel, R. B. Numerical methods for unconstrained optimization and nonlinear equations. Society for Industrial and Applied Mathematics. 1996.
11. Russell, S. J., & Norvig, P. Artificial intelligence a modern approach. London.2010.

# Basic Concepts of Optimization 2

Chapter 1 provided a fast introduction to optimization. The objective was to introduce the main concepts with a light reading to show how funny and exciting this field is, but optimization is also a wide field, and other concepts must be introduced before reviewing the algorithms in a more formal way.

Continuing with the initial concepts, this chapter will discuss some important facts about optimization. There are several topics that are important to know before starting coding optimization algorithms. The intention of this chapter is for the reader to know them, although we will refine them more in future chapters. The main topics covered in this chapter are (i) the zero derivative method, (ii) how to estimate the gradient numerically, (iii) graphical functions, (iv) situations in which the gradient descent algorithm may fail, and (v) the introduction to the random search algorithm, which is the first probability-based method.

## 2.1 The Zero-Derivative Method

It is exciting to see how the descendent gradient method can solve Johns's problem and the farmer fence problem. We would like to pause to address a topic that might cause concern after explaining the aforementioned problems and their solutions with the descendent gradient method in Matlab.

It is appropriate to briefly discuss the zero derivative method, especially if the reader is familiar with differential calculus or has taken engineering courses. You might notice that in calculus courses, a very quick and efficient optimization method is taught: the zero derivative method.

In the previous chapter, we studied Johns's problem, a problem in which we need to minimize the following objective function (This expression was used in Code 1.1).

$$f(x) = \frac{50}{x} + \frac{0.05}{2}x + 5. \tag{2.1}$$

Note that the derivative of Eq. (2.1) with respect to $x$, which is the gradient, can be expressed as (2.2) (This expression was also used in Code 1.1).

$$\frac{d}{dx}f(x) = -\frac{50}{x^2} + 0.025. \tag{2.2}$$

It's possible to set the derivative equal to zero and solve for $x$, the result of this is that $x = 44.7213$. This method is simple and effective and will obviously be the first option in cases where the derivative is known and where the decision variable(s) can be easily solved. However, this is not possible for many optimization problems. In fact, this will be discussed later when discussing a modification of John's problem.

Moreover, even if the derivative is defined, it could have several zeros, meaning several points where it crosses zero. Thus, the zero derivative method could become complicated. The problems in Chapter 1 were introduced so that the reader could understand the principle of the gradient descent method. Then, we will use it in problems in which the zero derivative method cannot be used.

The ultimate goal of this book is the application of numerical methods in optimization problems, including those that cannot be solved with the zero derivative method or the gradient descent method. There are various reasons why the objective function may be non-differentiable, for example, if it is discontinuous or if it is given by a data table, so there will be particular points where the derivative equals infinity or negative infinity.

Learning to distinguish when we can use a certain method or when it is more convenient to use a certain method is an important part of optimization. For now, we can summarize that if the derivative is defined and/or can be easily calculated, the gradient descent method will be a good option, but we can also try the zero derivative method.

## 2.2 Calculating the Gradient When It's not Defined

There are situations where the gradient (the derivative) is not defined, but it can be calculated numerically. In this case, we can approximate it using Eq. (2.3).

$$\nabla f(x_n) \approx \frac{f(x_n + h) - f(x_n)}{h}. \tag{2.3}$$

Equation (2.3) can be explained with Fig. 2.1. The gradient can be approximated as the ratio of the increase in the function divided by the increase in the independent variable. In

## 2.2 Calculating the Gradient When It's not Defined

**Fig. 2.1** How to obtain an approximation to the derivative

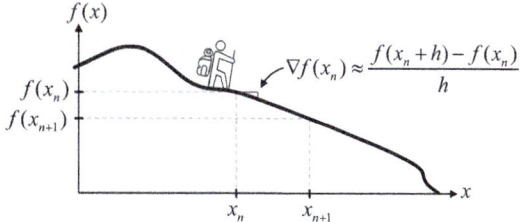

fact, the definition of a derivative is this equation when the increment of the independent variable tends to zero.

Code 2.1 solves Johns's problem, but instead of using the gradient expression, it applies Eq. (2.3) to approximate the derivative numerically:

```
% Code 2.1 - Problem 1.0, gradient (calculated) descent
clear ; clc ; close ;   % reset

f = @(x) 50./x + 0.05*x./2 + 5 ;    % Function

% Optimizer parameters
max_iter = 200 ;      % maximum number of iterations
q = 20 ;              % starting point
alpha = 100 ;         % learning rate
h = 0.001 ;           % delta of q

for i = 1:max_iter   % iterations
    grad = ( f(q+h) - f(q) )/( (q+h) - q ) ;
    q = q - alpha * grad ;
end

Solution = q
```

The main difference between Codes 1.2 and 1.1 is that in the Code 1.2, an equation for the gradient is not defined, but it is calculated using the objective function's equation and Eq. (2.3), which, in this case, is invoked in the part of the program that performs the iterations. Note that the increment of q is defined as:

$$h = 0.001 ;$$

This increment tends to be zero in the derivative definition. In the `for` loop that performs the algorithm, the line before the calculation of the new q, calculates the gradient with the formula (2.3):

$$grad = ( f(q+h) - f(q) )/( (q+h) - q ) ;$$

Upon executing the code, it yields the result of 44.7209, very close, although not identical to Code 1.1. Still, with this result, we can advise John to buy 45 gallons of water each time he goes to town.

## 2.3 The Stopping Criterion

So far, we have programmed a fixed number of iterations to perform. We can also use a stopping criterion strategy where we can check if the gradient is small enough or if the change in the solution is small enough, for example, a thousandth of a millionth ($1 \times 10^{-9}$). This way, we can consider that the derivative or the change in the solution is so small, practically zero, considering the algorithm has actually reached the solution. And we can stop the algorithm, determining if the program stopped because of the stopping criterion, or if it did because it reached the maximum number of iterations. In this case, we do not know if the gradient was close enough to zero to consider the result obtained as a good solution.

Code 2.2 incorporates this function. Again, it solves John's problem, but it has a stopping criterion.

```
% Code 2.2 - Problem 1.0, gradient descent, calculating the
% gradient and with early stopping criterion

clear ; clc ; close ;   % reset

f = @(x) 50./x + 0.05*x./2 + 5 ;        % Function

% Optimizer parameters
max_iter = 200 ;    % maximum number of iterations
q = 20 ;            % starting point
alpha = 100 ;       % learning rate
h = 0.001 ;         % delta of q
tol = 1e-9;         % tolerance for stopping strategy

for i = 1:max_iter   % iterations
    grad = ( f(q+h) - f(q) )/( (q+h) - q ) ;
    q = q - alpha * grad ;

    if abs(grad) < tol    % Checks convergence
        Conver_itera = i
            break
    end

end

Solution = q
```

If we execute the code, we get the result shown in Fig. 2.2, which indicates the solution but also tells us in which iteration the solution was achieved. In this case, the code did

## 2.4 Grapich Functions: Plotting the Function and the Current Solution

**Fig. 2.2** Result with the stopping strategy

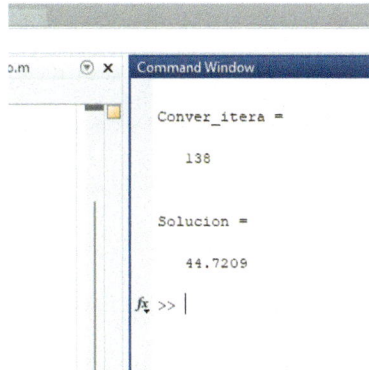

not reach iteration 200 because the gradient was smaller than a thousandth of a millionth at iteration 138.

Note that at the end of the definition of the optimizer parameters, a tolerance for the stopping criterion is defined as:

```
tol = 1e-9;  % tolerance for stopping strategy
```

The stopping strategy is within the `for` loop, where we ask with an `if` condition if the absolute value of the gradient is smaller than the tolerance. Only if this is true will the code within the loop be executed.

```
if abs(grad) < tol        % Checks convergence
    Conver_itera = i
    break
end
```

If the condition is met, the code displays the current iteration (`Conver_itera = i`) in the command window, which allows us to see in which iteration convergence was achieved. This process is called convergence, which is said to be when the program or algorithm converges. After this, the instruction ("`break`") stops the iterations of the for loop so it does not continue executing.

## 2.4 Grapich Functions: Plotting the Function and the Current Solution

We will now add some plots that will help us understand what is happening. Internally, Code 2.3 is similar to Code 2.2, but it includes some plotting functions that will be explained. But first, let's proceed to execute the following code.

```
% Code 2.3 - Problem 1.0, gradient descent, calculating the
% gradient, with stopping strategy, and plotting
clear ; clc ; close ;    % reset

f = @(x) 50./x + 0.05*x./2 + 5 ;       % Function

% Plotting the objective function
x = [ 1 : 1 : 200 ] ;    % generating an axis
f_x = f(x);              % evaluate the function
plot(f_x)                % plot
hold on ;                % indicates we don't want to replace the graph
axis ([ 0 200 6 16]) ;   % adjust the zoom

% Optimizer parameters
max_iter = 200 ;         % maximum number of iterations
q = 20 ;                 % starting point
alpha = 100 ;            % learning rate
h = 0.001 ;              % delta of q
tol = 1e-9;              % tolerance for stopping strategy

% Before we start, we draw the initial point
plot(q, f(q), 'ro', 'MarkerSize', 2 ) ;  % draws a point

for i = 1:max_iter   % iterations
    grad = ( f(q+h) - f(q) )/( (q+h) - q ) ;
    q = q - alpha * grad ;

    if abs(grad) < tol       % Checks convergence
        Conver_itera = i
        break
    end

    plot(q, f(q), 'ro', 'MarkerSize', 2 ) ;
    pause(0.1);      % pause for 0.1 seconds to see it

end

Solution = q
```

Upon executing the code, we can verify that the solution is the same as in the previous, but now it shows an image with a convex graph to which points are added as the code progresses until the last point is positioned in the valley of the graph, see Fig. 2.3.

The reader is invited to change some parameters, for example, the starting point (`q = 20 ; % starting point`), the learning rate (`alpha = 100 ; % learning rate`), the tolerance (`tol = 1e-9; % tolerance for stopping strategy`), the increment (`h = 0.001 ; % delta of q`),, etc., to appreciate the changes in the algorithm.

We can see that the points are the solutions that the algorithm is exploring, and they are drawn on a previously arranged graph. In Code 2.3, below the definition of the objective function, the function is initially plotted with the following code.

## 2.4 Grapich Functions: Plotting the Function and the Current Solution

**Fig. 2.3** Graph of the result and the points explored

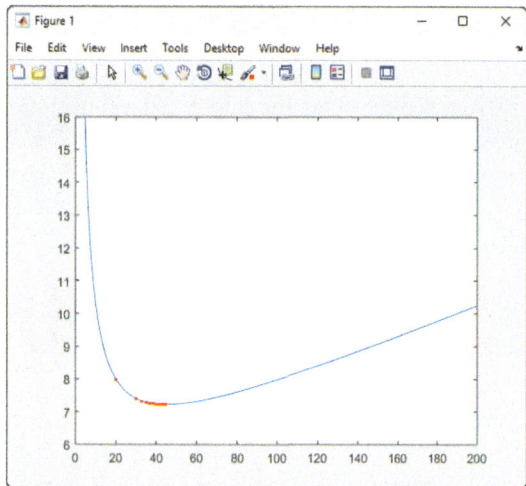

```
% Plotting the objective function
x = [ 1 : 1 : 200 ] ;    % generating an axis
f_x = f(x);              % evaluate the function
plot(f_x)                % plot
hold on ;                % indicates we don't want to replace the graph
axis ([ 0 200 6 16]) ;   % adjust the zoom
```

As indicated, to make the plot, we must define all the points of the $x$-axis that we want to evaluate. John's car has a maximum capacity of 200, and then the graph is plotted up to $q = 200$. The first line (x = [ 1: 1: 200];) generates a vector called x of 200 elements that starts with 1 and increments by one up to 200.

The second line [which evaluates the function; (f_x = f(x);)] generates another vector of 200 elements in which the values of evaluating the function $[f(1) f(2) \ldots f(200)]$ are stored, this second vector is called f_x. Then, we plot the function (plot(f_x)) and afterward adjust the axes with a convenient zoom to appreciate what happens with the objective function (axis ([ 0 200 6 16])).

The instruction hold on; (between the plot and the axis instructions) tells Matlab that all the following draws (basically points in this case) will be drawn over the existing graph. Otherwise, Matlab would erase the previous graph to draw the points as a new graph, this allows the point to be over the graph like in Fig. 2.3.

Before starting the algorithm's iterations, we draw f(q) the initial point with the following instructions:

```
% Before we start, we draw the initial point
plot(q, f(q), 'ro', 'MarkerSize', 2 ) ; % draws a point
```

The last change in Code 2.3 compared to 2.2 is that after each iteration, below the code that checks for convergence, the newly obtained point is drawn (over the graph).

```
plot(q, f(q), 'ro', 'MarkerSize', 2 ) ;
pause(0.1);      % pause for 0.1 seconds to see it
```

Note that the computer might perform operations so quickly that we could not visually appreciate what is happening internally. To visually appreciate how the new points are being drawn and how the algorithm is exploring the search area, we pause the computer for 100 ms after each iteration. This last line is optional, and the duration of the pause is also at the user's discretion.

## 2.5 Plotting Partial Costs and the Total Cost

Now that we know how to draw graphics let us step back for a moment to see how each of the costs behaves, which will help us understand the problem's behavior and provide some definitions. Below is Code 2.4, which plots the costs that eventually add up in the objective function.

```
% Code 2.4 - Problem 1.0, cost graphs

clear ; clc ; close ;    % reset

% Problem parameters
d = 5 ;          % daily gallon demand
k = 10 ;         % cost to go to town for the gallons
h = 0.05 ;       % daily storage cost per gallon
P = 1 ;          % unit cost of the gallons

x = [ 1 : 1 : 200 ] ;     % generating an x axis (q)
CTr = k.*d./x ;           % Average transportation cost
CAl = x.*h./2 ;           % Average storage cost
CPr = d*P*ones(1,200) ;   % Average product cost (gallons)

CTot = CTr + CAl + CPr ;

plot(x, CTr, x, CAl, x, CPr, x, CTot, 'linewidth', 2')    % plot
axis ([ 0 200 0 12]) ;    % adjust the zoom
```

If everything goes well, Code 2.4 should display the graph shown in Fig. 2.4, which shows the average transportation costs (in blue), storage costs (in orange), the unit cost of the products (the gallons of water) (in yellow), and the total cost (in purple). These costs are plotted against $Q$, that is, against the number of gallons John buys when he goes to town.

Minimizing the total average cost is the optimization goal of this application. We can notice that there are two costs influencing the location of the total cost's minimum point: the transportation cost and the storage cost.

## 2.5 Plotting Partial Costs and the Total Cost

**Fig. 2.4** Cost graphs without volume discounts

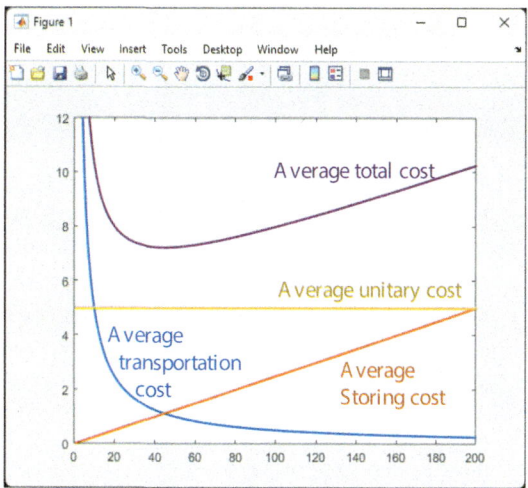

The average transportation cost, calculated with Eq. (1.2), is very high when John buys a few gallons and very low when $Q$ is high. Remember, it's an average cost; the cost of going to town is fixed ($k = 10$), but if many gallons are bought in one trip, many days can pass without making that expense. On the other hand, the storage cost increases. This is because if $Q$ is very large, some gallons will spend much more time in storage, paying the daily 5 cents per gallon for storage. The cost of the gallons doesn't change, so the yellow line is a straight line. The total cost, which is the purple line, is a convex curve that has a unique valley.

Some details to observe in Code 2.4 are the following.

The problem parameters are declared at the beginning of the program; in the previous codes, they were implicit in the definition of the objective function.

```
% Problem parameters
d = 5 ;           % daily gallon demand
k = 10 ;          % cost to go to town for the gallons
h = 0.05 ;        % daily storage cost per gallon
P = 1 ;           % unit cost of the gallons
```

Declaring them and using them later for calculations is a more elegant programming style and is more convenient if a change is desired. For example, if one of the costs changes, it is enough to modify its respective parameter.

A function (@(x)) that can be evaluated for various values is not defined. However, the cost operations are done vectorially, which is why the dots appear before the multiplication and division signs, indicating to Matlab that the elements used are vectors. The same applies to matrices; in fact, Matlab identifies our variables as $1 \times 200$ matrices.

```
x = [ 1 : 1 : 200 ] ;       % generating an x axis (q)
CTr = k.*d./x ;             % Average transportation cost
CAl = x.*h./2 ;             % Average storage cost
CPr = d*P*ones(1,200) ;     % Average product cost (gallons)

CTot = CTr + CAl + CPr ;
```

In the case of the average product cost, which is a constant scalar (a $1 \times 1$ matrix), we couldn't add it to the vectors, but by multiplying it by a $1 \times 200$ matrix of ones (`ones(1,200);`), it becomes a vector made of identical numbers.

Finally, to put several curves on the same graph, it is necessary to repeat in each curve what will be its independent variable. In our example, in all cases, this will be the x, which is the vector containing the different values that the variable $Q$ can take.

```
plot(x, CTr, x, CAl, x, CPr, x, CTot, 'linewidth', 2')      % plot
axis ([ 0 200 0 12]) ;   % adjust the zoom
```

Additionally, at the end of the plot command, we can make some modifications to the graphs, such as, in this case, making the line width 2. The last line adjusts the axes to a zoom that's convenient for viewing.

The gradient descent method is one of the most utilized optimization algorithms, but sometimes it doesn´t work, and then we have to use a different algorithm. Let us discuss the main reason for that effect.

## 2.6 When the Gradient Descent Method Does not Work

Revisiting the concepts learned from the gradient method, now that we have defined a strategy for "descending the hill," which can be summarized as always moving downwards, we can divide hills into two types: those that can be descended with this rule and those that cannot. Figure 2.5 shows a situation where always moving downwards does not ensure that we can reach the base; on the contrary, always moving downwards will stop us in a small valley that exists in the hill of Fig. 2.5. We will call this valley the local optimum, and the existence of these local optima in the objective functions poses interesting challenges in the field of optimization.

## 2.6 When the Gradient Descent Method Does not Work

**Fig. 2.5** A hill that cannot be descended with the gradient descent method

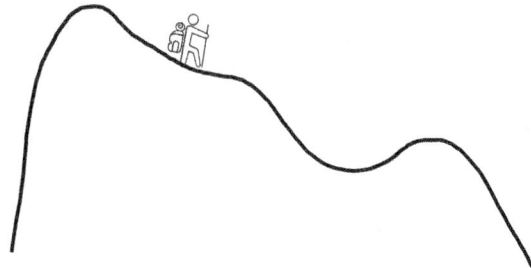

Some real-life situations can introduce local optima in the objective function that is desired to be optimized, as will be seen in the following example, which considers John's problem.

After solving his optimization problem, John began buying 45 gallons and visiting the town every nine days. Although he was aware that when a parameter change occurred, he would have to perform the optimization again, he now had tools like the Matlab codes and the gradient descent method.

One fine day, his supplier decided to implement a strategy of volume discounts, something very common in all markets; suppliers offer certain discounts to those who decide to buy a considerable volume of merchandise.

John's supplier announced that if he bought 120 or more gallons in a single purchase, instead of charging him 1 USD per gallon, he would charge 75 cents per gallon, and if he bought 160 gallons or more, he would give them to him at 60 cents each. The question now is whether he should change the amount of gallons he buys, that is, if he should optimize his objective function again.

The problem has a slight change, and volume discounts are very common in the economy; however, that slight change causes problems for the gradient descent method. Code 2.5 attempts to solve the problem now posed with the gradient descent method.

```
% Code 2.5 - John's Problem with volume discounts, with the
% gradient descent algorithm, without plotting.
clear ; clc ; close ;    % reset

% Problem parameters
d = 5 ;             % daily gallon demand
k = 10 ;            % cost to go to town for the gallons
h = 0.05 ;          % daily storage cost per gallon
%P = 1 ;            % unit cost of the gallons

Px(   1:  119) = 1 ;     % Cost with quantity discounts
Px( 120:  159) = 0.75 ;
Px( 160:  200) = 0.60 ;

x   = [ 1 : 1 : 200 ] ;  % generating an x axis (q)
CTr = k.*d./x ;          % Average transportation cost
CAl = x.*h./2 ;          % Average storage cost
CPr = d*Px ;             % Average product cost (gallons)

f = CTr + CAl + CPr ;

% Optimizer parameters
max_iter = 5000 ;   % maximum number of iterations
q        = 20 ;     % starting point
alpha = 5 ;         % learning rate
h     = 1 ;         % delta of q

for i = 1:max_iter    % iterations
    grad = ( f(round(q+h)) - f(round(q)) )/( (q+h) - q ) ;
    q = q - alpha * grad ;

end

Solution = q
```

Before we look at the solutions provided, let's take a closer look at the code, especially the new parts. For example, after defining the problem parameters, we notice that the unit cost is commented out (it has the %), and then we define the cost in parts:

```
% Problem parameters
d = 5 ;             % daily gallon demand
k = 10 ;            % cost to go to town for the gallons
h = 0.05 ;          % daily storage cost per gallon
%P = 1 ;            % unit cost of the gallons

Px(   1:  119) = 1 ;     % Cost with quantity discounts
Px( 120:  159) = 0.75 ;
Px( 160:  200) = 0.60 ;
```

This is how you construct a vector of 200 values in Matlab, but from 1 to 119, it has values of 1; from 120 to 159, it has values of 0.75; and from 160 to 200, it has values of 0.6.

## 2.6 When the Gradient Descent Method Does not Work

Subsequently, we need to calculate the costs. The calculation looks a bit different because now we want to use vectors to do it. The first line of the following code generates a vector of 200 units that will be used as a reference, both for the calculations and for the graphs. Afterward, the average costs are calculated, the transportation cost (formula (1.2)), the storage cost (formula (1.4)), and the product cost (formula (1.5)).

```
x   = [ 1 : 1 : 200 ] ;   % generating an x axis (q)
CTr = k.*d./x ;           % Average transportation cost
CAl = x.*h./2 ;           % Average storage cost
CPr = d*Px ;              % Average product cost (gallons)

f = CTr + CAl + CPr ;
```

Finally, these costs are added up, and the objective function is stored in the vector "f". After that, we define the optimizer parameters and apply the gradient descent algorithm.

However, when calculating the gradient, since the cost is now defined in a vector whose independent variable is an integer (x), we need to round the value obtained with the formulas, which is what the instruction ((round(q)) does:

```
for i = 1:max_iter   % iterations
    grad = ( f(round(q+h)) - f(round(q)) )/( (q+h) - q ) ;
    q = q - alpha * grad ;

end
```

When we run the code, it indicates the previous solution, although it provides us with the solution with decimals (for example, 44.5), we can easily evaluate the function at 44 and 45 ($f(44)$ and $f(45)$) in the command window, to find that the solution without volume discounts is 45. However, the solution is no longer 45 but 120. We can notice this if we type $f(120)$ in the command window and compare this average cost with the previously obtained average costs.

Even if we initialize the algorithm with a number greater than $q = 120$, it is unlikely to give us the result, and there could be some starting points that end in a program error.

Code 2.5 has trouble finding the solution; the main problem is that, with the introduction of volume discounts, the objective function became discontinuous, and in the field of optimization, we can say it is non-differentiable. This term is commonly heard when talking about numerical optimization algorithms since problems, where the objective function is non-differentiable, are where non-gradient-based algorithms find their main applications.

Code 2.6 graphs the objective function along with the average costs so that we can visually appreciate the discontinuous nature of the objective function.

```
% Code 2.6 - Problem 1.0, cost graphs with volume discounts

clear ; clc ; close ;    % reset

% Problem parameters
d = 5 ;            % daily gallon demand
k = 10 ;           % cost to go to town for the gallons
h = 0.05 ;         % daily storage cost per gallon

P(   1:  119) = 1     ; % Cost with quantity discounts
P( 120:  159) = 0.75 ;
P( 160:  200) = 0.60 ;

x = [ 1 : 1 : 200 ] ;   % generating an x axis (q)
CTr = k.*d./x ;         % Average transportation cost
CAl = x.*h./2 ;         % Average storage cost
CPr = d*P ;             % Average product cost (gallons)

CTot = CTr + CAl + CPr ;

plot(x, CTr, x, CAl, x, CPr, x, CTot, 'linewidth', 2')      % plot
axis ([ 0 200 0 12]) ;  % adjust the zoom
```

When we run this code, we can see the graph in Fig. 2.6. It looks similar to 1.13, but the discontinuity in the unit cost and its effect on the total cost, which is the objective function, can be observed.

If the algorithm tries to calculate the gradient at a point of discontinuity, the value obtained is so large that it can cause an error, for example, suggesting that the next value of $q$ is 500 (when $q$ is bounded between 1 and 200). The most complicated issue is that the new global optimum is precisely at a point of discontinuity (120).

**Fig. 2.6** Cost graphs with volume discounts

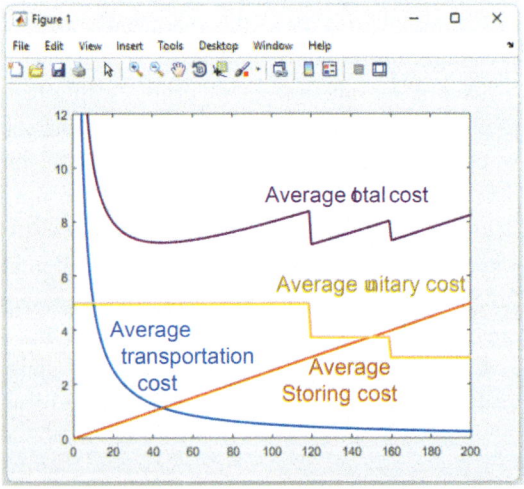

Returning to the issue of the solution, the new solution is to buy 120 gallons at a time, but how can we make the computer find this value now? Let's be patient; we are just getting started, but indeed, this concern will be resolved in this chapter.

The gradient descent method is not capable of solving the problem now that volume discounts have been introduced, as these discounts introduce discontinuities to the objective function that make it non-differentiable. We will solve this problem with some stochastic iterative methods, starting with the simplest, the Random Search.

## 2.7 The Random Search Algorithm

We'll start with one of the simplest and most elementary methods. Despite its simplicity, it is powerful and contains some elements of the optimization methods that will be addressed in this book.

The random search method [1], as its name suggests, involves randomly searching within the solution search space. Solutions generated by this method are evaluated against the objective function and under any constraints (if any). A variable in the program's memory stores the best solution found during the iterations, which is updated throughout the iterations if better solutions are found. This process repeats until the maximum number of iterations previously programmed is completed.

Although this approach may seem simple, it offers an effective alternative for optimizing very complex problems [2]. Some examples of its applications are parameter selection in machine learning models or in high-dimensionality situations. In other words, our problem is relatively simple considering the type of problems [3–6] that are usually solved with random iterative methods. We just have to find the value of one variable: how many bottles of water John should buy each time he goes to town.

The random search method [7] is particularly useful in situations with limited computational resources.

The random search method would consist of randomly selecting a value between 1 and 200, then evaluating the solution considering Eq. (1.6), repeated here as (2.4).

$$f(Q) = \frac{d}{Q}k + \frac{Q}{2}h + dP. \tag{2.4}$$

Remember that $d = 5$ gallons per day, $k = 10$ USD per trip to town, and $h = 0.5$ USD per gallon per day, but $P$ depends on the solution to be tested ($Q$). It is $P = 1$ USD per gallon if $120 < Q < 160$ and equals 0.60 if $160 \leq Q$.

Once the objective function is evaluated, in the first iteration, it is stored, and in subsequent iterations, the solution is compared against the stored one. The best solution is kept until a pre-set number of iterations is reached.

At first glance, it seems like an inefficient method since we have to repeat the process many times to find the solution. The good news is that a computer can do this very quickly; the same computer that was programmed with the gradient descent method could not complete the task.

On the other hand, we must remember that we will approach this method as an introduction to other methods that also have a component of randomness but with much more advanced strategies. These methods have been used to solve very complex problems, problems that conventional mathematics was unable to solve.

To execute the random search procedure, we can follow these steps:

(i) Define a domain of possible values for each solution parameter. This domain can be continuous (like a range of decimal numbers) or discrete (like a list of options or integers).

In our case, the solution has a single parameter, which is the number of gallons to buy, and it is a discrete parameter, as the purchase of fractions of a gallon is not contemplated but whole gallons. The range can go from 1 to 200.

(b) Set the maximum number of iterations that will be executed in the method, i.e., the total number of combinations that will be tested. This can be adjusted considering computational time and/or a limit of resources. However, as we discussed, a modern computer can perform hundreds or thousands of iterations in the time it would take a person to perform one iteration.

In the random search used for this case, the search is performed uniformly across the defined search space, meaning each combination has the same probability of being selected. Regardless of where the best solution is located within the search space, there will be other methods that do consider information from previous iterations to refine the search. These methods will be covered in later chapters.

The random search method is simple and efficient in terms of computational resources compared to other search methods. We could say it's a powerful optimization technique in its simplicity. Moreover, as will be seen in later chapters, it has various variants, such as local random search and adaptive random search, with modifications that improve some aspects of the original algorithm.

Code 2.7 solves John's problem with volume discounts using the random search method.

## 2.7 The Random Search Algorithm

```
% Code 2.7 - John's problem, with volume discounts, solved
% using the random search method.
clear ; clc ; close ;   % reset

% Problem parameters
d = 5 ;             % daily gallon demand
k = 10 ;            % cost to go to town for the gallons
h = 0.05 ;          % daily storage cost per gallon
P = 1 ;             % unit cost of the gallons

Px(   1: 119) = 1 ;        % Cost with quantity discounts
Px( 120: 159) = 0.75 ;
Px( 160: 200) = 0.60 ;

x = [ 1 : 1 : 200 ] ;   % generating an x axis (q)
CTr = k.*d./x ;         % Average transportation cost
CAl = x.*h./2 ;         % Average storage cost
CPr = d*Px ;            % Average product cost (gallons)

f = CTr + CAl + CPr ;

% Optimizer parameters
max_iter = 400 ;    % maximum number of iterations
q = 20 ;            % starting point
n = 0 ;             % improvement counter

for i = 1:max_iter  % iterations

    q_n = 1 + round(199*rand) ; % generating a random step

    if ( f(q_n) < f(q)) % evaluating if an update is needed
        q = q_n ;   % updates

        n = n + 1 ; % increments the improvement counter
        [ n i q ]   % displays the solution and counters

    end

end

Solution = q
```

The main parts that distinguish this algorithm are located within the for loop of the iterations. The function (rand) generates a random value between 0 and 1. We need a value between 1 and 200, so if we multiply the value of (rand) by 200, the random value would be between 0 and 200. However, we don't need to evaluate the value of 0, as the volume of purchase cannot be zero. Therefore, we multiply it by 199 and add one, ensuring that the randomly generated value, which could be the new solution to test, is between 1 and 200.

```
q_n = 1 + round(199*rand) ; % generating a random step
```

This new solution is stored in a variable (q_n). After that, we have an `if` statement. In this case, we ask if the new randomly generated solution is better than the best one tested so far, which has been stored in a variable (q). If it is the first iteration, the variable q contains the starting point, which in Code 2.7 is equal to 20. After the first iteration, the variable q is only updated if a better solution is found.

In fact, the if statement could be as simple as the following:

```
if ( f(q_n) < f(q) )   % evaluating if an update is needed
    q = q_n ;          % updates
end
```

Obviously, the variable q will contain, in the end, the best of the 400 solutions tested. This number was stored in the variable max_iter before starting the for loop. Although the if statement could be simple, Code 2.7 includes a couple of lines that serve to display information.

```
if ( f(q_n) < f(q) )   % evaluating if an update is needed
    q = q_n ;          % updates

    n = n + 1 ;        % increments the improvement counter
    [ n i q ]          % displays the solution and counters

end
```

The variable (n), called the improvement counter and initialized with the value of zero, will count how many times the code inside the (`if`) was entered, i.e., 400 randomly found solutions will be evaluated. However, the code inside the if is executed only when the new solution found is the best found so far, so each time it is entered, the improvement counter (n) is incremented, and then a vector with three variables is constructed. This is a simple and practical way to show three values on the same command window line.

This vector ([ n i q]) shows the improvement counter (n), i.e., how many times a better solution has been found. It also shows the for loop counter (i), just to know in which of the 400 iterations the current best solution was found, and finally the updated solution (q), which is the current best solution.

Running the code, we notice that it correctly solves the problem in a very short time, compared to the gradient descent method, which had errors. It's impressive how such a simple algorithm can solve the problem with high speed and precision. Almost always, the result is the global optimum. If we increase the number of iterations, reliability significantly increases, and the correct result is shown almost all the time.

As we advance through the chapters, we'll cover other algorithms more rigorously. We'll also tackle more complex, multimodal, and multidimensional problems solved with methods capable of providing solutions to problems with an infinite amount of possible solutions. This is just the tip of the iceberg, only a small sample of an extensive field of research, which has decades of development but remains a relatively new field that has recently evolved and gained momentum with the growing interest in artificial intelligence.

Finally, before going depper with the optimization methods, let us introduce a couple of key concepts in the difficulty of solving a problem: multimodality and multidimensionality.

## 2.8 Modality and Dimensionality

The first of these terms was mentioned in John's problem and the problem of descending a hill. After explaining the gradient method, we mentioned that we could now divide hills into two types: those that can be descended (or ascended) with the help of the gradient method, which consists of always going downhill (or uphill), and the hills where this method is not effective, see Fig. 2.7.

In the context of optimization, a problem is multimodal when it has multiple optimal local solutions, the best of which is, by definition, the global optima. These different local maxima or minima are called *local optima*, and when there are multiple *local optima* in a problem, we say it is a multimodal problem [8].

Another way to specify that a problem is unimodal is to mention that its objective function is convex. The presence of multiple *local optima* can complicate the task of finding the global maximum or minimum of a function, especially when using gradient-based optimization methods. This is because these methods can get trapped in a local maximum or minimum and not reach the global one. This happened to John's problem and usually occurs in the EOQ problem when volume discounts are introduced.

To address multimodal problems, more sophisticated optimization techniques are often used, which will be covered in this book, such as genetic algorithms, particle swarm optimization, and other evolutionary methods [9], which are less prone to getting stuck in local optima and have a higher probability of finding the global optimum in multimodal problems.

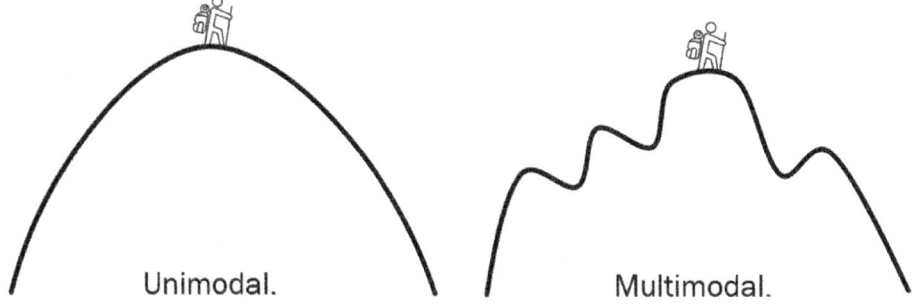

**Fig. 2.7** Types of hills according to their modes (unimodal and multimodal)

**Fig. 2.8** Two-dimensional hill, one-dimensional problem

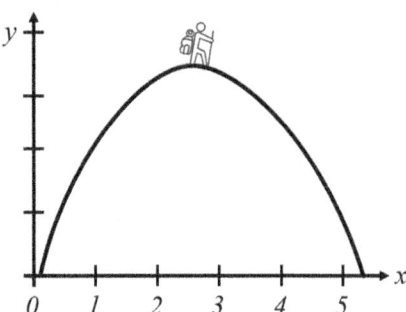

Let us know to explain the *dimensionality* of the problem. The dimension of a problem refers to the number of independent variables that must be considered to find a solution. Essentially, it is a measure of how complex a problem is from the perspective of the search space.

In a two-dimensional graph, finding a certain point requires two variables. Let's assume they are $(x, y)$. The optimization problem would consist of having one in the function of the other $(y = f(x))$, and we could maximize or minimize $y = f(x)$ by choosing $x$. Figure 2.8 shows a two-dimensional hill; the problem of finding the peak is one-dimensional because there is one independent variable (x); the other is not independent but depends on the first, $y = f(x)$.

Figure 2.9 shows a hill in three dimensions; in this case, the maximization problem is two-dimensional because there are only two independent variables and one dependent.

Note that instead of using the names $(x, y)$ for the axes, we can use the notation $(x_1, x_2)$, which is more common in the field of optimization.

The maximum height of the hill is 50 m, and the height depends on where we are located. For example, at the origin ($x_1 = 0$, $x_2 = 0$), the height $y = f(x_1, x_2)$ equals zero, and the maximum is found at the point ($x_1 = 5$, $x_2 = 5$) with a height $y = f(x_1, x_2) = 50$ (meters).

In the following chapters, it will be explained that the EOQ problem, John's problem, which was addressed as one-dimensional, commonly becomes multidimensional with another phenomenon that is also very common in the industry, occurring when suppliers do not have enough production capacity to meet demand, then the demand must be covered with a combination of purchases from various suppliers.

The dimension of the problem has important implications in terms of the difficulty of solving it [10]. As the dimension increases, the space of possible solutions can grow significantly. This phenomenon is colloquially known as the "curse of dimensionality." As the number of dimensions increases, more sophisticated algorithms and/or more computational time are required to explore and find optimal solutions in that expanded space.

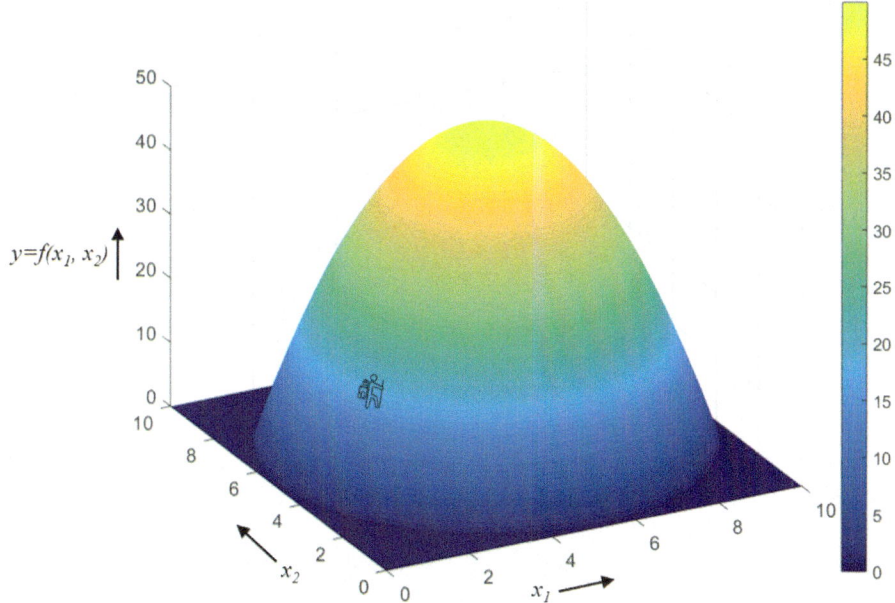

**Fig. 2.9** Three-dimensional hill, two-dimensional problem

This is especially true for exhaustive search methods, where the number of possible combinations can grow rapidly with each additional dimension. Therefore, high-dimensional problems often require special techniques or approximations to be solved effectively.

## 2.9 Homework Example—Maximizing the Force Among Charges with the Gradient Descent Method

We usually describe atoms as units composed of sub-particles: protons, neutrons, and electrons, often saying there is an equal number of protons and electrons. This is true for atoms in equilibrium or without an electric charge, but atoms can be imbalanced in the number of protons and electrons. This imbalance leads to the existence of electric charge and electricity. For example, in batteries, the negative pole has atoms with more electrons than protons, and the positive pole has the opposite. When a circuit is closed, electrons from atoms with excess electrons travel to balance the atoms with fewer electrons. When all atoms are balanced, the battery is fully discharged.

Electrons move driven by the force of attraction between electric charges, an invisible force of nature, somewhat similar to the force of gravity, but gravity is caused by the mass (or weight) of atoms, whereas the force of attraction is caused by electric charge, which is that imbalance between electrons and protons. It seems nature wants to be balanced

**Fig. 2.10** Force of attraction between charges of different signs

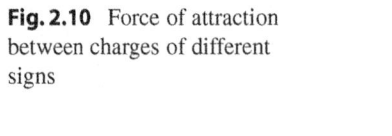
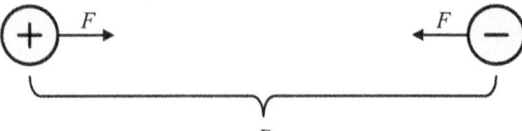

**Fig. 2.11** Force of repulsion between charges of the same sign

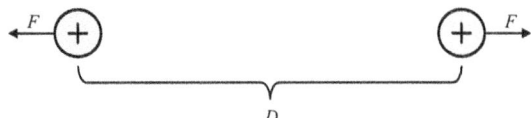

and is willing to perform work towards that end, and we, as humanity, have learned to control this invisible and wonderful force, using it in our devices and gadgets.

In the field of Electricity and Magnetism, the force of attraction between charges can be calculated with an equation called Coulomb's law. The force can be attractive if the particles have charges of different signs; for example, see Fig. 2.10. It doesn't matter which charge is on which side; they are just different. Both charges feel a force of equal magnitude but in opposite directions, pushing them to unite.

The force can be repulsive if the charges have the same sign, see Fig. 2.11. It doesn't matter what the signs are, but they are the same sign (both positive or negative).

The force is proportional to the product of the charges and inversely proportional to the square of the distance between the particles. Coulomb's law equation is as follows (2.5).

$$F = K \frac{Q_1 Q_2}{D^2}. \tag{2.5}$$

where $K$ is Coulomb's constant $K = 1/4\pi\varepsilon 0 = 8.9875 \times 10^9$ N.m$^2$/C$^2$. The unit of charge is Coulomb, the unit of distance is meters, and the unit of force is Newtons. Coulomb's constant is usually rounded to $9 \times 10^9$.

Suppose you have a total charge $Q_T = 100 \times 10^{-6}$ C (one hundred micro coulombs) that you can use to charge two particles. For this, we must distribute the charge into two parts, $Q_1$ and $Q_2$, being both numbers positive so that (2.6) is fulfilled.

$$Q_1 + Q_2 = Q_T. \tag{2.6}$$

If the particles are separated by a distance of 0.15 m (fifteen centimeters), find the charge distribution, i.e., what $Q_1$ and $Q_2$ should be, so that (2.6) is fulfilled and also maximize the repulsion force between them.

To solve this problem, we can use the gradient method, in this case, ascent, since it's a maximization problem. However, first, we must adapt the model to the canonical form of a maximization problem. For this, we can express $Q_2$ in terms of $Q_T$ and $Q_1$, see (2.7).

## 2.9 Homework Example—Maximizing the Force Among Charges ...

$$Q_2 = Q_T - Q_1. \tag{2.7}$$

And substitute (2.7) into Coulomb's law (2.5), obtaining.

$$F = K \frac{Q_1(Q_T - Q_1)}{D^2}. \tag{2.8}$$

This allows us to have the objective function in terms of a single variable and use the gradient method to solve it.

The problem can be expressed as follows:

$$\text{Max } F = K \frac{Q_1(Q_T - Q_1)}{D^2}. \tag{2.9}$$

Subject to:

$$Q_1 + Q_2 = Q_T. \tag{2.10}$$

$$Q_1 > 0. \tag{2.11}$$

$$Q_2 > 0. \tag{2.12}$$

And with the following parameters $K = 9 \times 10^9$, $Q_T = 100 \times 10^{-6}$, $D = 0.15$.
Code 2.8 solves the described problem.

```
% Code 2.8 - maximize force between charges
clear; clc; close;   % reset

C = 100e-6 ;
K = 9e9 ;
D = 0.15 ;

f = @(x) K*( x*C - x^2 )/D^2 ;      % Function
%grad_f = @(x) (K/d^2)*(C-2*x) ;     % Gradient

% Optimizer parameters
max_iter = 200 ;       % maximum number of iterations
q1 = 10e-6 ;           % starting point
alpha = 1e-12 ;        % learning rate
h = 1e-12 ;            % delta for q1

for i = 1:max_iter    % iterations
    grad = ( f(q1+h) - f(q1) )/h ;
    q1 = q1 + alpha * grad ;
end

Solution = q1
```

Running the code in Matlab, it is noticeable that the solution is to have $Q_1 = 50 \times 10^{-6}$, and consequently $Q_2 = 50 \times 10^{-6}$. Any other combination of charges would have a lesser repulsion force.

If you've reached this point, you will now see the enormous potential mastering these topics would have, the power to solve the mysteries of optimization, which have a vast field of applications in both professional and personal realms. Let us continue.

## References

1. Russell, S. J., & Norvig, P. (2010). Artificial intelligence a modern approach. London.
2. Kaelo, O. (2006). Some Variants of the controlled random search algorithm for global optimization. Journal of optimization theory and appplications. 130 (2), 253-264.
3. Price, W.L. (1987).Global optimization algorithms for a CAD workstation. Journal of optimization theory and applications. 55(1), 133-146.
4. Thakur, G., Pal, A., Mittal, N., Yajid, M.S.A., Gared, F. (2024). A significant exploration on meta-heuristic based approaches for optimization in the waste management route problems. Scientific reports. 14(1).
5. Cui, E.A., Zhang, Z., Chen, C.J., Wong, W,K. (2024).Applications of nature-inspired meta-heuristic algorithms for tackling optimization problems across disciplines. Scientific reports. 14(1).
6. Rezk, H., Ghani, O.A., Wilberforce, T., Taha, S.E. (2024). Metaheuristic optimization algorithms for real-world electrical and civil engineering application: A review. Results in Engineering. 23.
7. Price, W.L. (1983).Global optimization by controlled random search. Journal of optimization theory and applications. 40(3), 33-348.
8. Ali, M.M., Storey,C. (1994). Modified controlled random search algorithms. International journal of computer mathematics. 53(3-4), 229-235.
9. Beltran, L.A., Navarro, M.a., Oliva, D., Campos-Peña, D., Ramos-Frutos, J.A., Zapotecas-Martínez, S. (2024). Quasi-random Fractal Search (QRFS): A dynamic metaheuristic with sigmoid population decrement for global optimization. Expert systems with Applications. 254 (15).
10. Rao, S. S., (2019). Engineering Optimization, Theory and Practice. Wiley.

# The Gradient Descent Method Generalization for N Dimensions

## 3.1 Introduction

As mentioned before, optimization is a natural activity and part of everyday life. But it is also an important activity for the industry. Most companies need to optimize various processes, such as inventory costs and production costs, in their supply chain. These activities can include purchasing materials, selecting suppliers, transportation, and manufacturing processes [1–3].

In Chap. 1, we provided an introduction to the book topic and the story of John; in Chap. 2, we provided some basic concepts. John´s problem is a simple version of what, in the field of engineering, is called the EOQ (Economic Order Quantity) problem [4, 5], a subject of study in Industrial Engineering. That chapter also introduced some specific topics and a couple of solution methods: the gradient descent and the random search, both reviewed in their simplest and most elemental form.

This chapter will explain the gradient descent method in its general form, which includes multidimensional problems.

## 3.2 The Gradient Descent Method

This method is one of the first algorithms used for optimizing objective functions of various dimensions. In many processes, the objective function is multidimensional, meaning the goal is to know the optimal value of several variables, which are represented in a decision vector, and the function can be non-linear. The requirement is that the mathematical function be differentiable since its operation is based on the calculation of the gradient or the use of the derivative. Due to its simplicity and ease of application, it is one of the most widely used methods [6, 7] and was first introduced by Augustin Louis Cauchy in

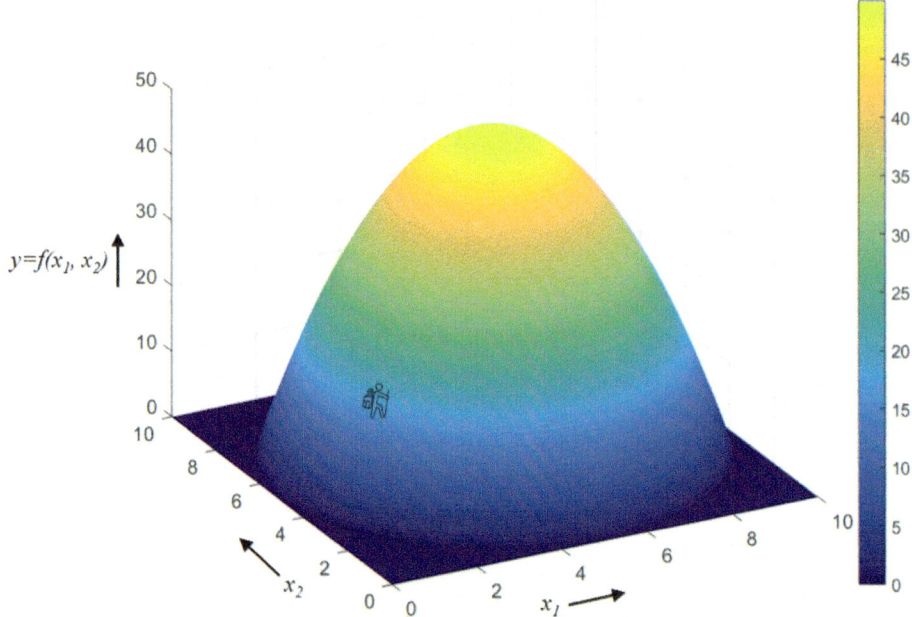

**Fig. 3.1** Three-dimensional hill, two-dimensional problem

the nineteenth century. This chapter will present the gradient method in its general form for a multidimensional problem, meaning that the solution is not a scalar but a vector. Subsequently, some problems will be reviewed.

We will start with the assumption that the objective function and the constraints, if any, are known. As an example, we will use the problem of climbing (maximizing) the 3D hill (see Fig. 3.1).

We can think at this time that the objective function of this 3D hill can be expressed as:

$$f(x_1, x_2) = 50 - (x_1 - 5)^2 - (x_2 - 5)^2. \tag{3.1}$$

As mentioned before, we could use the axes $(x, y, z)$, commonly used in some physics and mathematics courses, but we will use the nomenclature $(x_1, x_2, x_3)$, which is more commonly used in the field of optimization. Familiarizing oneself with this notation allows for easier extension of concepts to $N$-dimensional problems.

An initial point or initial solution is also required. This point can be selected randomly, or the programmer can propose one based on their experience.

In this chapter, and also in this book, vectors will be used once in a while; let us explain the notation. Vector will be written in bold, for example, the solution **x**, meaning the coordinates of a point, is a vector composed of two scalars ($x_1$ and $x_2$). Note that bold

## 3.2 The Gradient Descent Method

notation will be used to refer to vectors (or matrices if applicable) to distinguish them from scalars. The initial solution will be called $\mathbf{x}_0$; after this initial solution, a solution $\mathbf{x}_1, \mathbf{x}_2, \mathbf{x}_3, \ldots$ and so on will be generated until the iterative process is completed.

Remember that each of these solutions ($\mathbf{x}_1, \mathbf{x}_2, \mathbf{x}_3, \ldots, \mathbf{x}_{\text{Iter}}$) consists of two values for a two-dimensional problem, three values for a three-dimensional problem, or N values for an N-dimensional problem.

$$\mathbf{x}_0 = \begin{bmatrix} x_1[0] \\ x_2[0] \\ \vdots \\ x_N[0] \end{bmatrix}. \qquad (3.2)$$

In the case of the 3D hill, which is a two-dimensional problem, the initial solution can be (only as an example) (3.3).

$$\mathbf{x}_0 = \begin{bmatrix} x_1[0] \\ x_2[0] \end{bmatrix} = \begin{bmatrix} 2 \\ 4 \end{bmatrix}. \qquad (3.3)$$

The coordinates ($x_1 = 2$ and $x_2 = 4$) are (approximately) where our climber is in Fig. 3.1. From the initial point $\mathbf{x}_0$, the next point $\mathbf{x}_1$ will be obtained, from this one, $\mathbf{x}_2$ will be obtained, and so forth until a number of iterations we will call Iter. This is known as an iterative process, a step-by-step process, where each step evolves toward the final solution, which is expected to be the optimal solution $\bar{\mathbf{x}}$.

The independent variables, in this case, the coordinates ($x_1$ and $x_2$), are called decision variables and are usually bounded, meaning they cannot take any value. In our problem from Fig. 3.1, we can recognize that both $x_1$ and $x_2$ take values from 0 to 10. This information is important because there is a maximum and minimum value for each decision variable. It is possible to define vectors with the maximum and minimum values of the decision variables as:

$$\mathbf{x}_{\min} = \begin{bmatrix} x_{1\,\min} \\ x_{2\,\min} \\ \vdots \\ x_{N\,\min} \end{bmatrix}. \qquad (3.4)$$

$$\mathbf{x}_{\max} = \begin{bmatrix} x_{1\,\max} \\ x_{2\,\max} \\ \vdots \\ x_{N\,\max} \end{bmatrix}. \qquad (3.5)$$

Which in the hill example, it's possible to define vectors with the maximum and minimum values of the decision variables.

As mentioned, the initial point can be generated randomly, this can be done as (3.6).

$$x_0 = ((x_{max} - x_{min}) \odot \mathbf{rand}) + x_{min}. \tag{3.6}$$

The word rand is a function in Matlab, and other optimization programs, that refers to random. This function provides a quasi-random number between 0 and 1. The fact that **rand** is in bold indicates that it is a vector in which each element was generated with random values, and the symbol of a circle with a dot inside is known in mathematics as the Hadamard product, meaning it is an element-wise multiplication. In Matlab, this is indicated by adding a dot before the multiplication sign between two vectors or matrices.

These initial values will be modified in each iteration, using the following Eq. (3.7).

$$x_{k+1} = x_k - \alpha \nabla f. \tag{3.7}$$

where $x_{k+1}$ is the solution or position we are constructing from the previous solution $x_k$ and the step we are going to take. The step is defined as $-\alpha$ for the case of minimization or $+\alpha$ for the case of maximization, multiplied by the gradient of the function $\nabla f$. The nabla operator ($\nabla$) indicates the gradient of the function ($f$), somewhat similar to the derivative, but the derivative generates a scalar function from a scalar function, and the gradient generates a vector from a scalar function. In other words, the objective function can be a scalar function like (3.1). In a general case, the gradient of a scalar function that has $N$ independent variables $f(x_1, x_2, \ldots x_N)$ is calculated as (3.8).

$$\nabla f(x_1, x_2, \ldots, x_N) = \begin{bmatrix} \frac{\partial}{\partial x_1} f(x_1, x_2, \ldots, x_N) \\ \frac{\partial}{\partial x_2} f(x_1, x_2, \ldots, x_N) \\ \vdots \\ \frac{\partial}{\partial x_N} f(x_1, x_2, \ldots, x_N) \end{bmatrix}. \tag{3.8}$$

For the example of the 3D hill shown in Fig. 3.1 and whose function is defined in (3.1), it would be calculated as:

$$\nabla f(x_1, x_2) = \nabla \left(50 - (x_1 - 5)^2 - (x_2 - 5)^2\right). \tag{3.9}$$

$$\nabla f(x_1, x_2) = \begin{bmatrix} \frac{\partial}{\partial x_1} f(x_1, x_2) \\ \frac{\partial f}{\partial x_2} f(x_1, x_2) \end{bmatrix} = \begin{bmatrix} -2(x_1 - 5) \\ -2(x_2 - 5) \end{bmatrix}. \tag{3.10}$$

The gradient of a function at a point $x_i$ represents the direction in which the function has its maximum growth. If we want to climb the hill, we must move in that direction; if we want to descend, we must move in the opposite direction, which is why the minus sign in (3.7) is used for a minimization objective, to descend the hill, but the same equation with a plus sign would lead us to climb the hill.

Calculating the gradient then represents how our function varies concerning one of the variables in the decision vector, that is, concerning one of its dimensions [8].

The gradient can be calculated deterministically with (3.8), but it can also be approximated numerically, as seen in John's example, for each dependent variable, with the same concept of the derivative calculated numerically. For a system with three independent variables $(x_1, x_2, x_3)$, this can be done as follows:

$$\frac{\partial}{\partial x_1} f(x_1, x_2, x_3) = \frac{f((x_1 + h), x_2, x_3) - f(x_1, x_2, x_3)}{h}. \tag{3.11}$$

$$\frac{\partial}{\partial x_2} f(x_1, x_2, x_3) = \frac{f(x_1, (x_2 + h), x_3) - f(x_1, x_2, x_3)}{h}. \tag{3.12}$$

$$\frac{\partial}{\partial x_3} f(x_1, x_2, x_3) = \frac{f(x_1, x_2, (x_3 + h)) - f(x_1, x_2, x_3)}{h}. \tag{3.13}$$

In a general case, for a scalar function of $N$ independent variables $(x_1, x_2, ..., x_i, ..., x_N)$, the partial derivative of the function with respect to the i-th independent variable $(x_i)$ could be approximated as:

$$\frac{\partial}{\partial x_i} f(x_1, x_2, \ldots, x_i, \ldots, x_N)$$
$$= \frac{f(x_1, x_2, \ldots, (x_i + h), \ldots, x_N) - f(x_1, x_2, \ldots, x_i, \ldots, x_N)}{h}. \tag{3.14}$$

## 3.3   Finding the Peak of the 3D Hill

The first example we will tackle is how to find the peak of the 3D hill shown in Fig. 3.1. But before addressing that topic, we will review how to make a 3D graph in Matlab and specifically how to make the graph of interest (the 3D hill).

Let us start by drawing the 3D Hill. There are various ways to draw a figure as described, although exhaustively reviewing Matlab programming is beyond the focus of this book, a particular method will be used. However, we will review a specific detail of Fig. 3.1.

The surface described by Eq. (3.1) is similar but not exactly to the one shown in Fig. 3.1. Code 3.1 shows a graph of Eq. (3.1).

```
% Code 3.1 - Draw a 3D hill without a floor
clear ; clc ; close ;  % reset

f = @(x1, x2) 50 -2*(x1-5).^2 -2*(x2-5).^2 ; % equation (3.1)

% Create a mesh of points in space
[x1, x2] = meshgrid(linspace(0, 10, 1000), linspace(0, 10, 1000));

z = f(x1, x2); % Evaluate the function at each point on the mesh

% Draw the 3D surface
surf(x1, x2, z, 'EdgeColor', 'none', 'FaceColor', 'interp') ;
colorbar ; % displays the color bar.
```

As a result of running this code, we see in Fig. 3.2, an image that is very similar but not identical to Fig. 3.1. This figure is known as an elliptical paraboloid. To achieve an aesthetic appearance similar to a hill, we added a floor by setting the negative parts of this paraboloid to zero.

To generate the 3D hill depicted in Fig. 3.1 (which includes the floor), Code 3.2 was used.

```
% Code 3.2 - Draw a hill in 3D with a floor
clear ; clc ; close ;   % reset

f = @(x1, x2) 50 -2*(x1-5).^2 -2*(x2-5).^2 ; % equation (3.1)

% Create a grid of points in space
[x1, x2] = meshgrid(linspace(0, 10, 1000), linspace(0, 10, 1000));

z = f(x1, x2); % Evaluate the function at each grid point

for i1=1:1000
    for i2=1:1000
        if z(i1,i2)<0
            z(i1,i2)=0; % Set negative values to zero (floor)
        end
    end
end



    % Draw the 3D surface
    surf(x1, x2, z, 'EdgeColor', 'none', 'FaceColor', 'interp') ;
    colorbar ; % display the color bar.
```

The nested for loop traverses the entire graph to set negative points to zero. We can observe that executing this code yields a figure identical to Fig. 3.1. Now, we can use Fig. 3.1 as a didactic example. However, we must consider that if we try to maximize the function with the gradient method but initialize the system on the floor, where the derivative is equal to zero, the algorithm would not find the peak. For this reason, we will stop using the graph from Fig. 3.1 despite its closer resemblance to a hill, and we will use

## 3.3 Finding the Peak of the 3D Hill

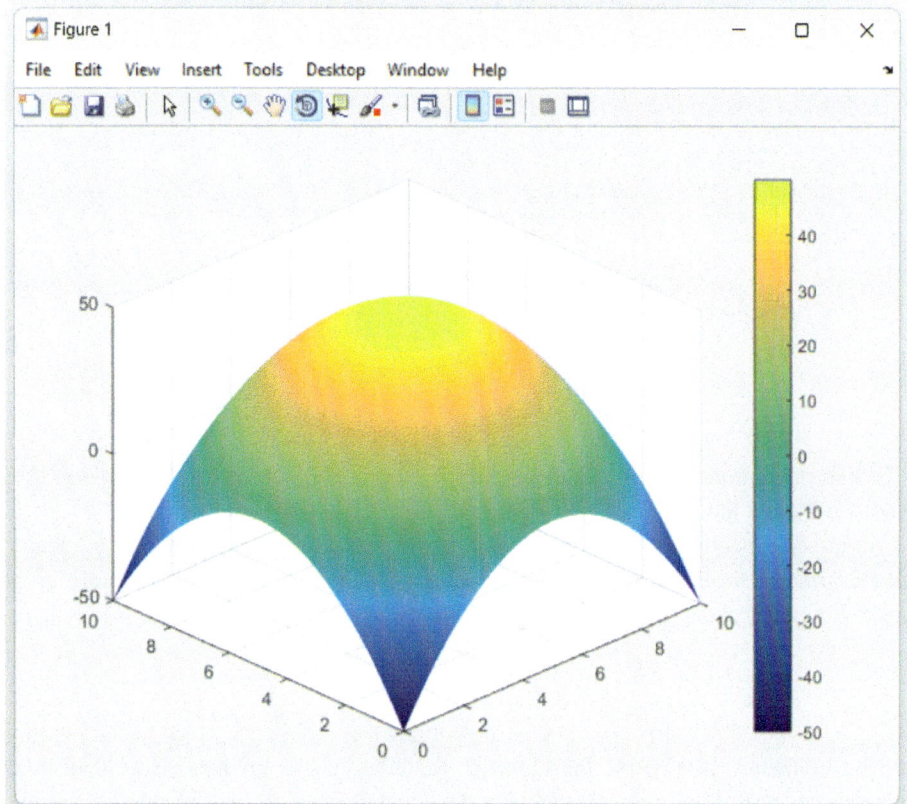

**Fig. 3.2** Elliptical paraboloid, a three-dimensional hill without a floor

the real elliptic paraboloid shown in Fig. 3.2. When we approach optimization techniques that do not depend on the gradient, we might return to the 3D hill with a floor.

Now that we know how to draw the hill, let us to maximizing the function. Code 3.3 applies the gradient technique to maximize the objective function described by the elliptic paraboloid given by Eq. (3.1) and shown in Fig. 3.2.

```
% Code 3.3 - optimize the 3D hill with the gradient method
clear ; clc ; close ;    % reset

% Definition of the function and its gradient
f      = @(x) 50 -2*(x(1)-5).^2 -2*(x(2)-5).^2 ; % equation (3.1)
grad_f = @(x) [ -2*(x(1)-5) ; -2*(x(2)-5) ] ;    % equation (3.11)

% Parameters for the gradient method
alpha = 0.1 ;   % Learning rate
Iter  = 1000 ;  % Maximum number of iterations
p = [ 1; 1]; % Starting point

for iter = 1:Iter % Iterative process
    p = p + alpha * grad_f(p) ; % Obtains next point
end

Sol = p           % Result coordinates of the highest point
Height = f(p)     % Height of the highest point
```

When running the code, we can see in the command window that the solution is the point (5, 5), and the height of the hill at that point is 50.

As already mentioned in the previous chapter, the first line resets Matlab's memory, clears variables, the command window, and closes auxiliary windows (for example, graphs made with other codes).

```
% Code 3.3 - optimize the 3D hill with the gradient method
clear ; clc ; close ;    % reset
```

The following lines define the objective function and its gradient. Note that, unlike previous codes, this code contains an important change; the functions receive an argument (**x**), but within them, access its components, $x_1$ and $x_2$ which in the code are (x(1) and x(2)), meaning **x** must be a 2-component vector. We don't need to state this explicitly, but if components of the argument are accessed within the function, then the function must receive a vector. We could make a function that receives a vector of any dimension, which could be useful if a high-dimensionality problem needs to be studied.

The same applies to the gradient function, but unlike the objective function, which receives a vector and returns a scalar, the gradient function receives a scalar function and returns a vector. As a result, it returns Eq. (3.10), which has two components. This is why brackets and the semicolon (;) are used within the function.

```
% Definition of the function and its gradient
f      = @(x) 50 -2*(x(1)-5).^2 -2*(x(2)-5).^2 ; % equation (3.1)
grad_f = @(x) [ -2*(x(1)-5) ; -2*(x(2)-5) ] ;    % equation (3.11)
```

These definitions simplify the operation of the optimization method, which, as we can observe, is very simple. The following lines set the parameters of the optimization method and the iterative process.

## 3.3 Finding the Peak of the 3D Hill

```matlab
            % Parameters for the gradient method
            alpha = 0.1 ;   % Learning rate
            Iter  = 1000 ;  % Maximum number of iterations
            p = [ 1; 1];    % Starting point

            for iter = 1:Iter % Iterative process
                p = p + alpha * grad_f(p) ;  % Obtains next point
            end

            Sol = p          % Result coordinates of the highest point
            Height = f(p)    % Height of the highest point
```

Code 3.4 performs the same optimization but includes the graph and places points where the optimizer is exploring the solution.

```matlab
        % Code 3.4 - optimize the hill in 3D with the gradient method
        clear ; clc ; close ;   % reset

        % Definition of the function and its gradient
        f_ = @(x1,x2) 50 -2*(x1  -5).^2 -2*(x2  -5).^2 ;   % equation (3.1)
        grad_f = @(x) [ -2*(x(1)-5) ; -2*(x(2)-5) ] ;      % equation (3.11)

        % Create a mesh of points in space
        [x1, x2] = meshgrid(linspace(0, 10, 1000), linspace(0, 10, 1000));
        z = f_(x1, x2); % Evaluate the function at each point on the mesh
        surf(x1, x2, z, 'EdgeColor', 'none', 'FaceColor', 'interp') ;
        colorbar ;  % displays the color bar.

        % Parameters for the gradient method


        alpha = 0.1 ;  % Learning rate
        Iter  = 100 ;  % Maximum number of iterations
        p = [ 0; 5] ;  % Starting point

        for iter = 1:Iter % Iterative process
            p = p + alpha * grad_f(p) ;  % Obtains next point

            hold on ;          % indicates that we do not replace the graph
            plot3(p(1), p(2), f_(p(1),p(2)), 'ro', 'MarkerSize', 2 );
            pause(0.05);       % we pause for 0.1 seconds to see it

        end

        Sol = p              % Result coordinates of the highest point
        Altura = f_(p(1),p(2))   % Height of the highest point
```

When running this code, we should get a graph similar to Fig. 3.2, but this time, red dots appear at the points where the algorithm searches for a new solution until it reaches the top.

The code has differences from 2.3 that will be discussed next, but it's worth noting that Matlab is a very powerful and flexible language, and almost all procedures can be done in various ways. We encourage the reader not to feel overwhelmed by the vast world of Matlab; it's not necessary to be an expert in Matlab to solve the problems we encounter, and the recommendation is to use the provided codes, especially those that best fit your style.

The main differences between this code and the previous one are:

The objective function f_ was defined to accept two arguments, each as a scalar. Remember that in code 2.3, the objective function received a vector composed of two scalars. The reason for this change is that this code will perform a graph of the function, and there's a method in Matlab to do this very simply, but the objective function must receive the arguments separately.

```
% Definition of the function and its gradient
f_ = @(x1,x2) 50 -2*(x1  -5).^2 -2*(x2  -5).^2 ;   % equation (3.1)
grad_f = @(x) [ -2*(x(1)-5) ; -2*(x(2)-5) ] ;      % equation (3.11)
```

Now, proceeding with the 3D graph, the meshgrid function generates two matrices, in this case of 1000 × 1000, which contain coordinates $(x, y)$ that could be the coordinates of each point on the hill. These coordinates can be used to evaluate the objective function, which would generate a third matrix ($z = f\_(\times 1, \times 2);$) with the values of the height of the hill. Now, we can use the surf function to draw the 3D surface.

```
% Create a mesh of points in space
[x1, x2] = meshgrid(linspace(0, 10, 1000), linspace(0, 10, 1000));
z = f_(x1, x2); % Evaluate the function at each point on the mesh
surf(x1, x2, z, 'EdgeColor', 'none', 'FaceColor', 'interp') ;
colorbar ; % displays the color bar.
```

The iterative process is performed in the same way, but within the for loop, a point is placed at the locations being explored with the following code.

Matlab

```
        for iter = 1:Iter % Iterative process
            p = p + alpha * grad_f(p) ; % Obtains next point

            hold on ;        % indicates that we do not replace the graph
            plot3(p(1), p(2), f_(p(1),p(2)), 'ro', 'MarkerSize', 2 );
            pause(0.05);     % we pause for 0.1 seconds to see it

        end
```

Code 3.5 performs the same function as Code 3.4, however, instead of defining the gradient as a function and evaluating it, it obtains the gradient numerically, which is done within the for loop of the iterative process. This book will present more than one programming style. Readers are invited to adopt the style that is most convenient for them.

## 3.4 Maximizing the Peaks Function

```
% Code 3.5 - optimize a hill in 3D - gradient method
clear ; clc ; close ;    % reset

% Definition of the function and its gradient
f = @(x1,x2) 50 -2*(x1  -5).^2 -2*(x2  -5).^2 ;  % equation (3.1)

% Create a grid of points in space
[x1, x2] = meshgrid(linspace(0, 10, 1000), linspace(0, 10, 1000));
z = f(x1, x2); % Evaluate the function at each point on the grid
surf(x1, x2, z, 'EdgeColor', 'none', 'FaceColor', 'interp') ;
colorbar ;      % displays the color bar.

% Parameters for the gradient method
alpha = 0.01 ;  % Learning rate
Iter  = 100  ;  % Maximum number of iterations
p = [ 0; 5]  ;  % Starting point
h = 0.001    ;

for iter = 1:Iter % Iterative process

    p(1) = p(1) + alpha * (f(p(1)+h,p(2)  ) - f(p(1),p(2)))/h ;
    p(2) = p(2) + alpha * (f(p(1)  ,p(2)+h) - f(p(1),p(2)))/h ;

    hold on ;      % indicates that we do not replace the graph
    plot3(p(1), p(2), f(p(1),p(2)), 'ro', 'MarkerSize', 2 );
    pause(0.05);    % we pause for 0.05 seconds to be able to see it

end

Sol = p          % Coordinates of the highest point as the result
Altura = f(p(1),p(2))    % Height of the highest point
```

### 3.4  Maximizing the Peaks Function

Let us try another multidimensional problem, but in this case, a multimodal one; let us try the Peaks function [9]. It is a function commonly used in numerical optimization courses. The Peaks function is described by Eq. (3.15).

$$f(x_i, x_2) = 3(1 - x_1)^2 \cdot e^{(-x_1^2 - (x_2+1)^2))} \\ - 10\left(\frac{x_1}{5} - x_1^3 - x_2^5\right) \cdot e^{(-x_1^2 - x_2^2)} - \frac{1}{3}e^{(-(x_1+1)^2 - x_2^2)}. \tag{3.15}$$

where the values of both $x_1$ and $x_2$ are within the interval $(-3 \leq x_1 \leq 3)$, $(-3 \leq x_2 \leq 3)$. Code 3.6 plots the Peaks function. Let us see the function before optimizing it,

```
% Code 3.6 - Plot the Peaks function in 3D
clear ; clc ; close ;    % reset

f = @(x1,x2) 3*(1-x1).^2.*exp(-(x1.^2) - (x2+1).^2) ...
    - 10*(x1/5 - x1.^3 - x2.^5).*exp(-x1.^2-x2.^2) ...
    - 1/3*exp(-(x1+1).^2 - x2.^2) ;

% Create a mesh of points in space
[x1, x2] = meshgrid(linspace(-3, 3, 1000), linspace(-3, 3, 1000)) ;
z = f(x1, x2) ; % Evaluate the function at each point on the mesh

figure('Position', [100 100 600 500]) ;
surf(x1, x2, z, 'EdgeColor', 'none', 'FaceColor', 'interp') ;
```

If everything goes well, running the code will display a figure like Fig. 3.3. Returning to the analogy of hills, the Peaks function would be a surface with three hills and two valleys. The valleys have a negative height, while the hills have a positive height, so the zero height would be like the ground level.

Among the tools that appear along with the figure is the *Rotate 3D* tool, whose icon is indicated with a red arrow in Fig. 3.3. Once we click and hover the mouse over the figure, we can see the cursor with a shape similar to the icon. Gently drag (press the mouse button and move the mouse without releasing the button) and observe how the 3D figure rotates, showing more details of its shape.

3D graphs help us visualize the evolution of the algorithm, but there is another way in 2D to show the peaks and valleys of a scalar field like this. It is the contour plot, and code 3.7 plots the Peaks function using contour plots.

**Fig. 3.3** Graph of the Peaks function in 3D

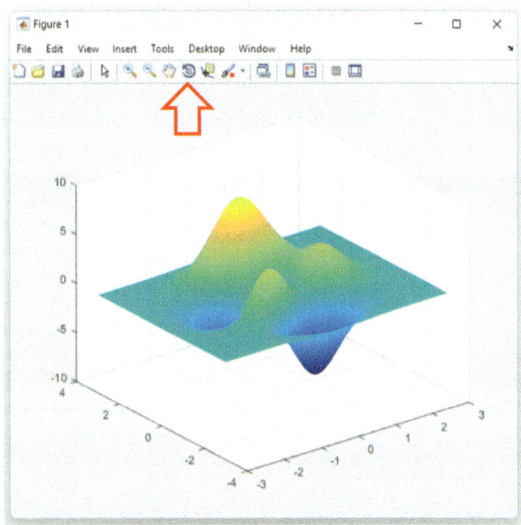

## 3.4 Maximizing the Peaks Function

**Fig. 3.4** Graph of the Peaks function in 2D (contours)

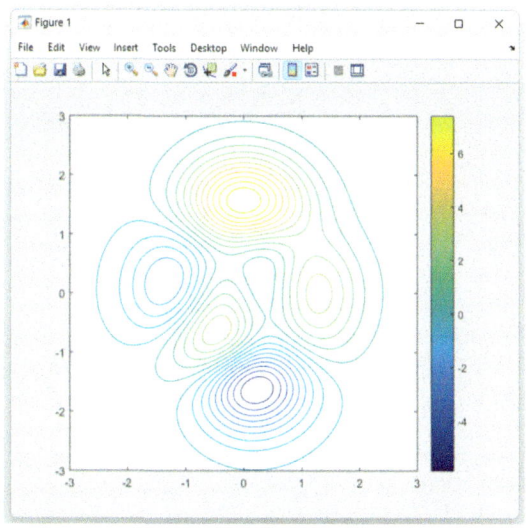

```
% Code 3.7 - Plot the Peaks function in 2D (contours)
clear ; clc ; close ;    % reset

f = @(x1,x2) 3*(1-x1).^2.*exp(-(x1.^2) - (x2+1).^2) ...
    - 10*(x1/5 - x1.^3 - x2.^5).*exp(-x1.^2-x2.^2) ...
    - 1/3*exp(-(x1+1).^2 - x2.^2) ;

% Create a mesh of points in space
[x1, x2] = meshgrid(linspace(-3, 3, 1000), linspace(-3, 3, 1000)) ;
z = f(x1, x2) ; % Evaluate the function at each point on the mesh

figure('Position', [100 100 600 500]) ;
contour(x1,x2,z,20) ;
colorbar ; % displays the color bar.
```

If everything goes well, running the code will display a figure like Fig. 3.4.

Returning to the analogy of hills, the color of the contours symbolizes the height of the contour. These contours are also used in the topography of real hills. The orange and yellow bars are the highest peaks, and the blue and purple circles are the deepest valleys. Observe Figs. 3.3 and 3.4, which will help to understand the peak function better.

Code 3.8 implements the gradient ascent algorithm to maximize the Peaks function.

```
% Code 3.8 - Optimize Peaks function with gradient
clear ; clc ; close ;   % reset

f = @(x1,x2) 3*(1-x1).^2.*exp(-(x1.^2) - (x2+1).^2) ...
    - 10*(x1/5 - x1.^3 - x2.^5).*exp(-x1.^2-x2.^2) ...
    - 1/3*exp(-(x1+1).^2 - x2.^2) ;

% Create a mesh of points in space
[x1, x2] = meshgrid(linspace(-3, 3, 1000), linspace(-3, 3, 1000)) ;
z = f(x1, x2) ; % Evaluate the function at each point on the mesh

figure('Position', [100 100 1200 400]) ;
subplot(1, 2, 1) ;
surf(x1, x2, z, 'EdgeColor', 'none', 'FaceColor', 'interp') ;
colorbar ; % displays the color bar.
hold on ;
subplot(1, 2, 2) ;
contour(x1,x2,z,20) ;


    % Parameters for the gradient method
    alpha = 0.01 ;   % Learning rate
    Iter  = 100 ; % Maximum number of iterations
    p = [ -1; 2] ; % Starting point
    h = 0.001 ;

    for iter = 1:Iter % Iterative process

        p(1) = p(1) + alpha * (f(p(1)+h,p(2)) - f(p(1),p(2)))/h ;
        p(2) = p(2) + alpha * (f(p(1),p(2)+h) - f(p(1),p(2)))/h ;

        hold on ;
        subplot(1, 2, 1) ;
        plot3(p(1), p(2), f(p(1),p(2)), 'ro', 'MarkerSize', 2 ) ;
        hold on ;
        subplot(1, 2, 2) ;
        plot(p(1),p(2),'.','markersize',10,'markerfacecolor','g') ;
        pause(0.05) ;

    end

    Sol = p          % Result coordinates of the highest point
    Height = f(p(1),p(2))    % Height of the highest point
```

If everything goes well, running the code will display Fig. 3.5, where the Peaks function in 3D can be seen, as well as a 2D representation, along with the points that the algorithm explores.

Now, we will go over some details of the code. In the first lines, the Peaks function is established, and the matrices or meshes $x_1$, $x_2$, and $z$ ($x_3$) are created, which is, in fact, the evaluation of the Peaks function.

## 3.4 Maximizing the Peaks Function

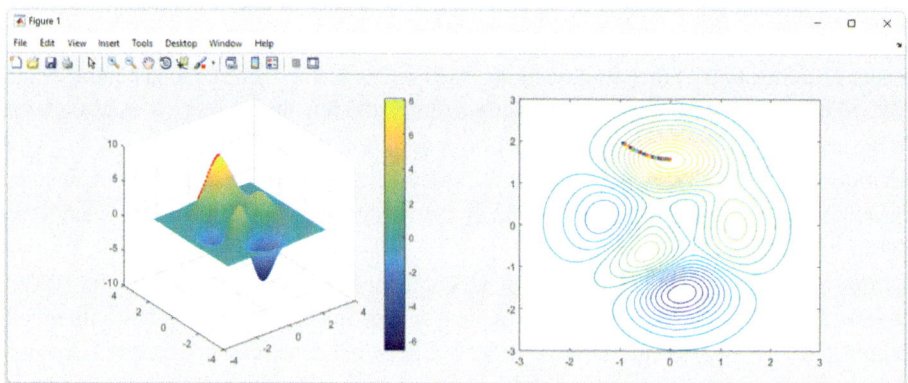

**Fig. 3.5** Graph of the gradient method's result on the Peaks function

```
% Code 3.8 - Optimize Peaks function with gradient
clear ; clc ; close ;   % reset
f = @(x1,x2) 3*(1-x1).^2.*exp(-(x1.^2) - (x2+1).^2) ...
    - 10*(x1/5 - x1.^3 - x2.^5).*exp(-x1.^2-x2.^2) ...
    - 1/3*exp(-(x1+1).^2 - x2.^2) ;

% Create a mesh of points in space
[x1, x2] = meshgrid(linspace(-3, 3, 1000), linspace(-3, 3, 1000)) ;
z = f(x1, x2) ; % Evaluate the function at each point on the mesh
```

The following lines create the figure. Note that now we use the command (figure('Position', [100 100 1200 400]);) it is not necessary to create the figure when one uses plotting functions like (surf, plot, contour), the figure is created automatically, but if we use this function we can consider some options, as in this case, we are specifying that the lower left corner is at the coordinates (100, 100) (of the screen), considering that the origin is the lower left corner of the screen. In addition, we are specifying that the width of the figure will be 1200 pixels and the height will be 400. The figure is relatively wide, and this is because we want to use two graphs, on the left side, the 3D graph, and on the right side, the contour representation, a way to graph in 2D a scalar function with two input arguments.

```
figure('Position', [100 100 1200 400]) ;
subplot(1, 2, 1) ;
surf(x1, x2, z, 'EdgeColor', 'none', 'FaceColor', 'interp') ;
colorbar ; % displays the color bar.
hold on ;
subplot(1, 2, 2) ;
contour(x1,x2,z,20) ;
```

The command (`subplot(1, 2, 1);`) indicates that within the created figure, not all space will be used, but the space will be divided into a $1 \times 2$ matrix (it can be more; the reader is invited to try other combinations, such as $2 \times 1$, $2 \times 2$, etc.), the last number indicates that the first space will be used. Subsequently, the function (`surf`) is used as before. In the second space, the `contour` function is used, `contour(x1, x2, z, 20)`, which performs the 2D graph shown on the right side of the figure, specifying the independent variables, the dependent variable, and finally, how many lines or contours we want to plot (20 in this case).

Subsequently, the parameters of the gradient method are declared, and the iterative process is performed. Note that the gradient of the functions is calculated numerically. Within the `for` loop of the iterative process, the same functions (`subplot`) are used, to indicate in which of the two graphs we want to put the point where the method is exploring.

```matlab
% Parameters for the gradient method
alpha = 0.01 ;   % Learning rate
Iter  = 100 ;    % Maximum number of iterations
p = [ -1; 2] ;   % Starting point
h = 0.001 ;


for iter = 1:Iter  % Iterative process

    p(1) = p(1) + alpha * (f(p(1)+h,p(2)) - f(p(1),p(2)))/h ;
    p(2) = p(2) + alpha * (f(p(1),p(2)+h) - f(p(1),p(2)))/h ;

    hold on ;
    subplot(1, 2, 1) ;
    plot3(p(1), p(2), f(p(1),p(2)), 'ro', 'MarkerSize', 2 ) ;
    hold on ;
    subplot(1, 2, 2) ;
    plot(p(1),p(2),'.','markersize',10,'markerfacecolor','g') ;
    pause(0.05) ;

end

Sol = p              % Result coordinates of the highest point
Height = f(p(1),p(2))   % Height of the highest point
```

At the end of code 3.8, the solution and its value in the objective function are shown. The Peaks function is widely used for testing optimization algorithms, so it is indeed predefined in Matlab, making it unnecessary to define it as was done in code 3.6. We can invoke it using the `peaks(x1, x2)` function.

## 3.4 Maximizing the Peaks Function

Code 3.9 performs the same operations as code 3.8, but instead of defining the Peaks function and calling (or invoking) it with `f(x1, x2)`, it uses the function defined in Matlab.

```matlab
% Code 3.9 - Optimize Peaks function with gradient
clear ; clc ; close ;     % reset

% Create a mesh of points in space
[x1, x2] = meshgrid(linspace(-3, 3, 1000), linspace(-3, 3, 1000)) ;
z = peaks(x1, x2) ; % Evaluate the function at each point on the mesh

figure('Position', [100 100 1200 400]) ;
subplot(1, 2, 1) ;
surf(x1, x2, z, 'EdgeColor', 'none', 'FaceColor', 'interp') ;
colorbar ; % displays the color bar.
hold on ;
subplot(1, 2, 2) ;
contour(x1,x2,z,20) ;

% Parameters for the gradient method
alpha = 0.01 ;   % Learning rate
Iter  = 100 ;  % Maximum number of iterations
p = [ -1; 2] ; % Starting point
h = 0.001 ;

for iter = 1:Iter % Iterative process

    p(1) = p(1) + alpha * (peaks(p(1)+h,p(2)) - peaks(p(1),p(2)))/h ;
    p(2) = p(2) + alpha * (peaks(p(1),p(2)+h) - peaks(p(1),p(2)))/h ;
    hold on ;
    subplot(1, 2, 1) ;
    plot3(p(1), p(2), peaks(p(1),p(2)), 'ro', 'MarkerSize', 2 ) ;
    hold on ;
    subplot(1, 2, 2) ;
    plot(p(1),p(2),'.','markersize',10,'markerfacecolor','g') ;
    pause(0.05) ;

end

Sol = p            % Result coordinates of the highest point
Height = peaks(p(1),p(2))    % Height of the highest point
```

Finally, in codes 3.8 and 3.9, the algorithm starts at $(-1, -2)$, which is the starting point. What happens if the starting point is different? Try running the code initializing the algorithm with another point, such as $(-0.5, -1)$. Doing this, we notice that the algorithm does not find the highest peak. The gradient algorithm will get trapped in a local optima. We invite the reader to try these and other options they find illustrative.

In the following chapters, metaheuristic methods such as adaptive random search or particle swarm optimization (PSO) will be used, and we will see how these algorithms based on probabilistic methods can find solutions under circumstances where gradient-based methods get trapped in local optima, regardless of the starting point of the algorithm.

## 3.5 Homework Example—Programming the Gradient Descent Method to Minimize the Bohachevsky Function

We will now try another multimodal and multidimensional function, which is also often used to test optimization algorithms. It's called the Bohachevsky function and is described by Eq. (3.16).

$$f(x_i, x_2) = x_1^2 + 2x_2^2 - 0.3\cos(3\pi x_1) - 0.4\cos(4\pi x_2) + 0.7. \tag{3.16}$$

This function has variants (with different parameters) that can further accentuate the multiple local optima, making its solution even more challenging. It is normally used as a test for minimization (not maximization).

The values of the independent variables ($x_1$ and $x_2$), are within the interval of $(-10 \leq x_1 \leq 10), (-10 \leq x_2 \leq 10)$.

*Answer*
Code 3.10 implements the gradient descent algorithm with the Bohachevsky function.

## 3.5 Homework Example—Programming the Gradient Descent ...

```matlab
% Code 3.10 - Optimize Bohachevsky function with gradient descent
clear ; clc ; close ;    % reset

f = @(x1,x2) (x1.^2) + 2.*(x2.^2)  -0.3.*cos(3.*pi.*x1) ...
             -0.4 .* cos(4.*pi.*x2) + 0.7 ;

% Create a mesh of points in space
[x1, x2] = meshgrid(linspace(-10, 10, 1000), linspace(-10, 10, 1000))
;
z = f(x1, x2) ; % Evaluate the function at each point on the mesh

figure('Position', [100 100 1200 400]) ;
subplot(1, 2, 1) ;
surf(x1, x2, z, 'EdgeColor', 'none', 'FaceColor', 'interp') ;
colorbar ; % displays the color bar.
hold on ;
subplot(1, 2, 2) ;
contour(x1,x2,z,20) ;

% Parameters for the gradient method
alpha = 0.02;   % Learning rate
Iter  = 100 ;   % Maximum number of iterations
p = [ -8; -8] ; % Starting point
h = 0.001 ;

for iter = 1:Iter % Iterative process

    p(1) = p(1) - alpha * (f(p(1)+h,p(2)) - f(p(1),p(2)))/h ;
    p(2) = p(2) - alpha * (f(p(1),p(2)+h) - f(p(1),p(2)))/h ;

    hold on ;
    subplot(1, 2, 1) ;
    plot3(p(1), p(2), f(p(1),p(2)), 'ro', 'MarkerSize', 2 ) ;
    hold on ;
    subplot(1, 2, 2) ;
    plot(p(1),p(2),'.','markersize',10,'markerfacecolor','g') ;
    pause(0.05) ;

end

Sol = p           % Result coordinates of the highest point
Height = f(p(1),p(2))   % Height of the highest point
```

If everything goes well, running the code will display Fig. 3.6, where the Bohachevsky function in 3D can be seen, as well as a 2D representation, along with the points that the algorithm explores.

The code is similar to the codes used for the Peaks function. The main difference is that, being a minimization problem, the equations calculating the next point within the iterative process have a negative sign, whereas previous codes had a positive sign.

```matlab
    p(1) = p(1) - alpha * (f(p(1)+h,p(2)) - f(p(1),p(2)))/h ;
    p(2) = p(2) - alpha * (f(p(1),p(2)+h) - f(p(1),p(2)))/h ;
```

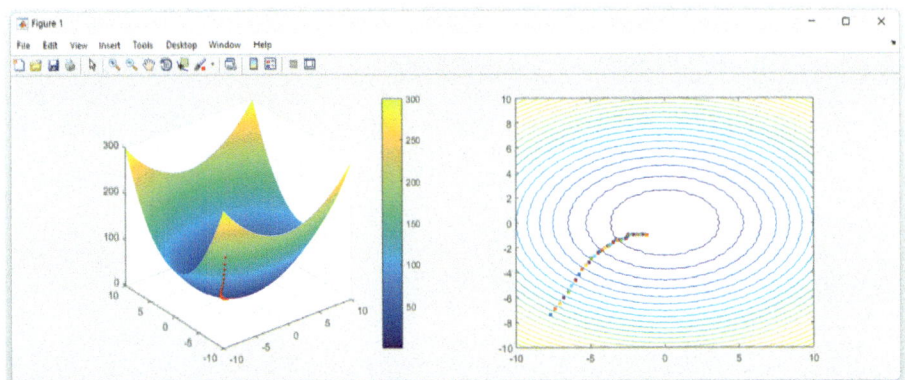

**Fig. 3.6** Graph of the gradient method's result on the Bohachevsky function

It's important to mention that the gradient method is not the most suitable for solving this type of multimodal and multidimensional problem. As mentioned in the example of the Peaks function, the solution obtained depended on the starting point.

In this case, the code cannot get the global optima. In other words, it cannot solve the problem since the solution is actually the point (0, 0). The reader is invited to try different parameters of the optimizer. The difficulty in minimizing the Bohachevsky function is due to the small (sinusoidal) waves that create local optima where the algorithm can get trapped.

The gradient descent algorithm is a powerful method in engineering, having proven its effectiveness in various problems. However, as we've seen in this chapter, in some cases, it just cannot find the answer.

In the following chapters, we will return to these types of problems, the Peaks function and the Bohachevsky function, with probabilistic methods, and we can have a clear idea of their benefits.

Let's move forward.

## References

1. Rao, S. S., Engineering optimization, theory and practice. Wiley, Fith edition, 2019.
2. Bonnans, J. F., Gilbert, J. C., Lemaréchal, C., & Sagastizábal, C. A. Numerical optimization: theoretical and practical aspects. Springer Science & Business Media, 2006.
3. Rsioshansi, Ramteen. Optimization in engineering: Models and Algorithms. SPRINGER, 2019.
4. Schwarz, L. B. The economic order-quantity (EOQ) model. Building Intuition: Insights from Basic Operations Management Models and Principles, 2008, 135–154.
5. Senthilnathan, S. Economic order quantity (EOQ). 2019, Available at SSRN 3475239.
6. Nocedal, J. Wright, S.J. Numerical optimization, Springer, 2006.
7. Goodfellow, I., Bengio, Y., and Courville, A. Deep learning. MIT press. 2016.

8. Mathews, J. H., & Fink, K. D. Numerical methods using MATLAB, Vol. 4. Upper Saddle River, NJ: Pearson prentice hall.2004.
9. Brero, A. C., & Gallard, R. H. Una comparación de algoritmos evolutivos para la optimización de funciones multimodales. In VI Congreso Argentino de Ciencias de la Computación.2000.

# Curve Fitting

## 4.1 Introduction to the Curve Fitting

Let's now look at an example whose mathematical content is used in various areas of engineering. The problem of curve fitting is multidimensional, as curves can be expressed by polynomials, but we can start with the simplest approach, which is a single-dimension problem [1, 2].

A railway constructor faces the following problem. It is desired to build a straight railway track that passes as close as possible to a group of cities, the map of which is shown in Fig. 4.1. The map places the city at the easternmost end in the center of a Cartesian plane where each unit is 100 km. This city will be the starting point of the railway, and the goal is to minimize the sum of the vertical distances (from north to south) of the other five cities to the track ($d_1 + d_2 + d_3 + d_4 + d_5$). The minimization will be achieved through the correct selection of the slope $m$ that will define the inclination of the tracks with respect to the horizontal (east to west line).

Remember that each point on a line that starts from the origin can be expressed with (4.1).

$$y = mx. \tag{4.1}$$

The coordinates of the cities can be seen in Table 4.1.

Thus, the sum of vertical distances, which can be the error or variable we want to minimize, can be written as (4.2).

$$\sum d = |1.5 - m| + |-0.5 - 2m|$$
$$+ |3.5 - 3m| + |3.5 - 4m| + |2.5 - 5m|. \tag{4.2}$$

**Fig. 4.1** Map of the cities

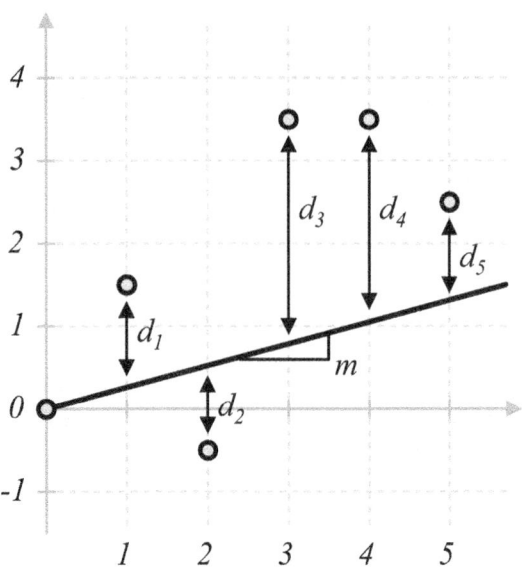

**Table 4.1** Coordinates of the citiesnadas de las ciudades

| x | 0 | 1 | 2 | 3 | 4 | 5 |
|---|---|---|---|---|---|---|
| y | 0 | 1.5 | − 0.5 | 3.5 | 3.5 | 2.5 |

Remember that each point on a line that starts from the origin can be expressed with (4.1). The absolute value is used instead of the simple difference because if the simple difference is used, a negative value could be canceled out by a positive value, making it seem like the sum of the distances is not decreasing when it indeed is.

Code 4.1 solves the proposed problem using the gradient descent method.

## 4.1 Introduction to the Curve Fitting

```
% Code 4.1 - Curve fitting (straight line) V1
clear; clc; close;   % reset


x = [ 0 , 1   ,  2  , 3  , 4  , 5  ];
y = [ 0 , 1.5 , -0.5 , 3.5 , 3.5 , 2.5 ];

m     = 0 ;      % Initial slope
alpha = 0.005 ;  % Learning rate
niter = 200 ;    % Number of iterations
h     = 0.001 ;  % delta of x_n

for i = 1:niter % Loop to perform gradient descent

    sum_d =   abs( y(2) -   m     ) + abs( y(3) - 2*m   ) ...
            + abs( y(4) - 3*m     ) + abs( y(5) - 4*m   ) ...
            + abs( y(6) - 5*m     ) ;

    sum_d_h = abs( y(2) -    (m+h) ) + abs( y(3) - 2*(m+h) ) ...
            + abs( y(4) - 3*(m+h) ) + abs( y(5) - 4*(m+h) ) ...
            + abs( y(6) - 5*(m+h) ) ;

    grad = (sum_d_h - sum_d ) / h ; % Calculates the gradient
    m = m - alpha * grad ;          % Updates the slope

end

Sol_m = m
```

If everything goes as planned, after solving the problem, Matlab will indicate that the solution is to have a slope $m = 0.86$. Note that the program runs very quickly, offering the solution almost instantaneously. The solution is a very good approximation. When applying an optimization method, there is a trade-off between precision and speed, which we will delve into later.

Note also that by using the numerical gradient calculation, we don´t need to know the derivative of Eq. (4.2) with respect to *m*, and still, we can solve the problem.

Code 4.2 presents the same program, with the difference being that it plots the location of the cities, as done in Fig. 4.1, plots the initial line with a slope of $m = 0$, and as the slope is updated, it plots one by one the updated lines until reaching its final version.

```matlab
% Code 4.2 - Curve fitting (straight line) V2 (with graph)
clear; clc; close;    % reset

x = [ 0 , 1   ,   2   , 3   , 4   , 5   ];
y = [ 0 , 1.5 , 1.5 , 3.5 , 3.5 , 5   ];

m     = 0 ;        % Initial slope
alpha = 0.005 ;    % Learning rate
niter = 200 ;      % Number of iterations
h     = 0.001 ;    % delta of x_n

scatter(x, y, 'b') ; % Plot the location of the cities
axis ([ 0 6 -6 6]) ; % adjust the zoom
hold on ; % Indicates dont erase graphs

for i = 1:niter % Algorithm loop

    sum_d =    abs( y(2) -   m   ) + abs( y(3) - 2*m   ) ...
             + abs( y(4) - 3*m   ) + abs( y(5) - 4*m   ) ...
             + abs( y(6) - 5*m   ) ;

    sum_d_h = abs( y(2) -   (m+h) ) + abs( y(3) - 2*(m+h) ) ...
            + abs( y(4) - 3*(m+h) ) + abs( y(5) - 4*(m+h) ) ...
            + abs( y(6) - 5*(m+h) ) ;

    grad = (sum_d_h - sum_d ) / h ; % Calculates the gradient
    m = m - (alpha * grad) ;        % Updates the slope

        plot(x, m.*x, 'r');         % Plots the new line
        pause(0.1);                 % Pauses for a moment

end

plot(x, m.*x, 'b');                 % Plots the new line
Sol_m = m
```

If everything goes well, the Matlab graph should look like in Fig. 4.2. The arrow indicates the direction in which the new lines are being drawn; we can observe that as the lines approach the solution, the distance between them becomes smaller.

Some comments we can add regarding this code are that, in the first lines, we declare the ordered pairs or the coordinates of the cities, which correspond to Table 4.1.

```matlab
x = [ 0, 1, 2, 3, 4, 5 ];
y = [ 0, 1.5, 1.5, 3.5, 3.5, 5 ];
```

## 4.1 Introduction to the Curve Fitting

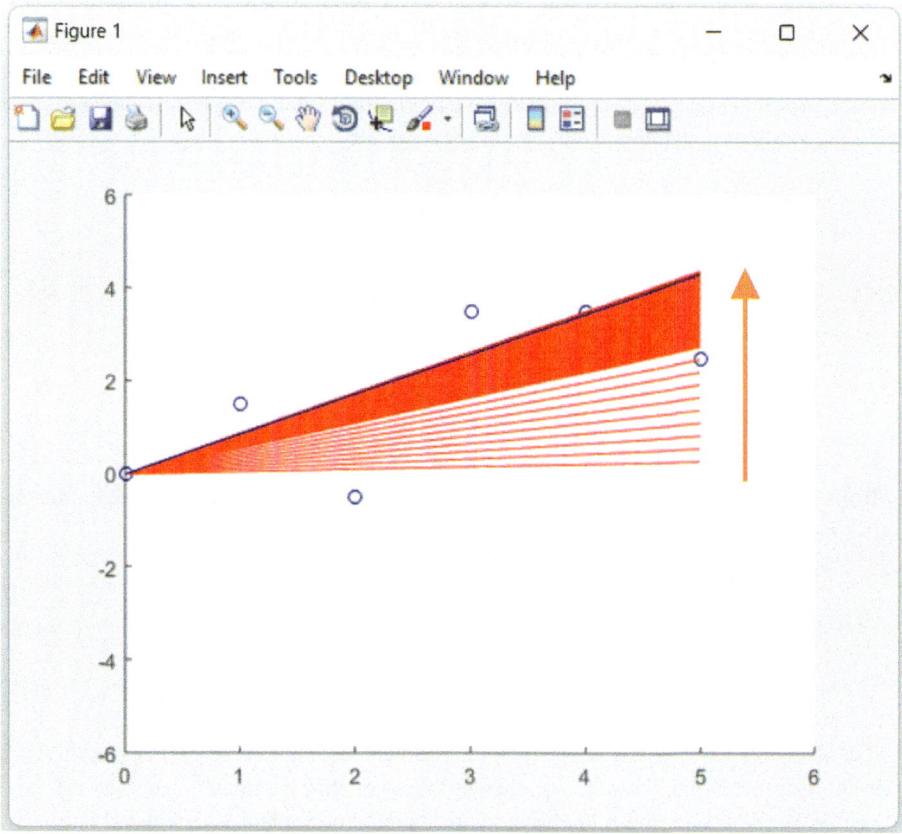

**Fig. 4.2** Graph of the result and the points that were explored

Subsequently, the optimizer parameters are declared before starting the iterations. The following lines reproduce Fig. 4.1, locating the cities on a Cartesian plane graph.

```
scatter(x, y, 'b') ; % Plot the location of the cities
axis ([ 0 6 -6 6]) ; % adjust the zoom
hold on ; % Indicates dont erase
```

Within the for loop that performs the steps of the optimization method, we can observe the calculation of the gradient, which is performed numerically. Initially, we calculate the distance with the current slope (m), then we calculate the distance with the current slope plus the increment (m + h), and finally, we apply the formula of the calculated gradient. Once the gradient is achieved, we update the slope *m*.

```
sum_d =    abs( y(2) -   m      ) + abs( y(3) - 2*m     ) ...
         + abs( y(4) - 3*m      ) + abs( y(5) - 4*m     ) ...
         + abs( y(6) - 5*m      ) ;

sum_d_h = abs( y(2) -    (m+h)  ) + abs( y(3) - 2*(m+h) ) ...
        + abs( y(4) - 3*(m+h)   ) + abs( y(5) - 4*(m+h) ) ...
        + abs( y(6) - 5*(m+h)   ) ;

grad = (sum_d_h - sum_d ) / h ;  % Calculates the gradient
m = m - (alpha * grad) ;         % Updates the slope
```

Before performing the next iteration, we graph the line with the updated slope and wait a moment to visually appreciate how the graphs are being updated.

```
plot(x, m.*x, 'r');              % Plots the new line
pause(0.1);                      % Pauses for a moment
```

Once the iterative process is finished, the program displays the solution on screen (Sol_m = m) and graphs one last straight line, in blue, in this case.

Try changing some points to see how the problem's solution changes. For example, try with the parameters of Table 4.2 or change the algorithm parameters, such as the initial slope, the learning rate, the h increment with which the gradient is calculated, etc.

Observe the behavior of the program with changes to its parameters.

Finally, before moving on to the next topic, Code 4.3 solves the problem using a more elegant form of programming, typically used when working with Matlab.

**Table 4.2** Alternative coordinates of the problem

| x | 0 | 1   | 2   | 3   | 4   | 5 |
|---|---|-----|-----|-----|-----|---|
| Y | 0 | 1.5 | 1.5 | 3.5 | 3.5 | 5 |

## 4.1 Introduction to the Curve Fitting

```matlab
% Code 4.3 - Curve fitting (straight line)
clear; clc; close;   % reset

x = [0, 1, 2, 3, 4, 5];
y = [0, 1.5, -0.5, 3.5, 3.5, 2.5];

m     = 0;        % Initial slope
alpha = 0.005;    % Learning rate
niter = 200;      % Number of iterations
h     = 0.001;    % delta of x_n

scatter(x, y, 'b'); % Graphs the location of the cities
axis([0 6 -6 6]);   % adjusts the zoom
hold on;  % Indicates don´t erase the draw

for i = 1:niter  % Algorithm loop

    sum_d   = sum(abs(y -  m   *x));
    sum_d_h = sum(abs(y - (m+h)*x));

    grad = (sum_d_h - sum_d) / h; % Calculates the gradient
    m = m - (alpha * grad);       % Updates the slope

    plot(x, m.*x, 'r');           % Graphs the new line
    pause(0.1);                   % Pauses for a moment

end

plot(x, m.*x, 'b');               % Graphs the new line
Sol_m = m
```

The calculations within the iteration cycle have been rewritten. Doing them point by point, as in the codes prior to this one, helps to understand what's happening inside the program. However, performing these operations with vectors (and eventually matrices) makes it more elegant. Moreover, if the number of points for curve fitting increases, the code doesn't change when using vector operations.

The lines of code used for the gradient calculation do exactly the same thing; they obtain the difference between the vector y and the straight line with slope $m$, take their absolute value, and then use the sum instruction to add up all the obtained values.

```matlab
sum_d = sum(abs(y - m *x));
```

This line of code could also be broken down into the nested operations to better understand the processes involved. The first operation performed is obtaining the error. This results in a vector of the same length as "y", in this case, six elements. Next, its absolute value is obtained, resulting in another vector of six elements, but now all positive. Finally, summing all the elements results in a scalar.

```
Error_vector = y - m*x;
Absolute_values = abs(Error_vector);
sum_d = sum(Absolute_values);
```

## 4.2 The Squared Error

Curve fitting, where we search for the parameters of a curve that make it resemble a series of previously described points, is a technique used in various engineering fields, sometimes referred to as linear regression [3, 4]. The goal is for the curve to pass through the points or as near as possible. The difference between the curve's value and the points is called the error. Until now, we have used the sum of errors in their absolute value.

It is more common to aim to minimize the squared error. The sum of the errors raised to the square. This approach avoids using the absolute value; by squaring the differences between the points and the plotted curve, negative differences become positive. Additionally, the squared error is differentiable, so we can calculate a derivative, which wasn't the case with the absolute value. This feature will be explained and used as an advantage. But before moving in that direction, let's look at code 4.4, which solves the curve fitting problem using the squared error.

## 4.2 The Squared Error

```
% Code 4.4 - Curve fitting (straight line - squared error)
clear; clc; close;   % reset

x = [ 0 , 1    ,  2   , 3    , 4   , 5    ] ;
y = [ 0 , 1.5  , -0.5 , 3.5  , 3.5 , 2.5 ] ;

m       = 0 ;          % Initial slope
alpha = 0.001 ;        % Learning rate
niter = 100 ;          % Number of iterations
h       = 0.001 ;      % delta for x_n

% Plot the original data
scatter(x, y, 'b');
axis ([ 0 6 -6 6]) ;   % adjust the zoom
hold on;

for i = 1:niter % Loop for gradient descent

    sum_d   = sum( (y -  m    *x).^2 ) ;
    sum_d_h = sum( (y - (m+h)*x).^2 ) ;

    grad = (sum_d_h - sum_d ) / h ;
    m = m - alpha * grad ;

    plot(x, m.*x, 'r');
    pause(0.1);

end

plot(x, m.*x, 'b');
Sol_m = m
```

If everything goes well, the program will show us a graph similar to the one in Fig. 4.3, and it will indicate that the result is 0.6813. It is normal for the result to differ when using squared error compared to using the absolute value of the error. In both cases, the algorithm tries to minimize a sum of errors, but with squared error, the algorithm attempts to "more significantly" minimize the larger errors, saying it "penalizes" the larger error values more.

To understand the differences between the absolute value and the squared error, let us discuss a little history.

Suppose we need to leave one place at 3:50 pm and arrive at another at 4:10 pm, but the travel time is 40 min. Inevitably, we will have an error. We could choose to leave the first appointment at 3:50 pm and arrive at the second appointment at 4:30 pm (20 min late), having an error of 0 min for the first activity and 20 min for the second. In absolute terms, it's the same as leaving the first appointment at 3:40 pm, incurring a 10-min error for the first activity and a 10-min error for the second. However, using squared error, the second option is better since the squared error would be $10^2 + 10^2 = 200$, which is lower than the other option, $0^2 + 20^2 = 400$.

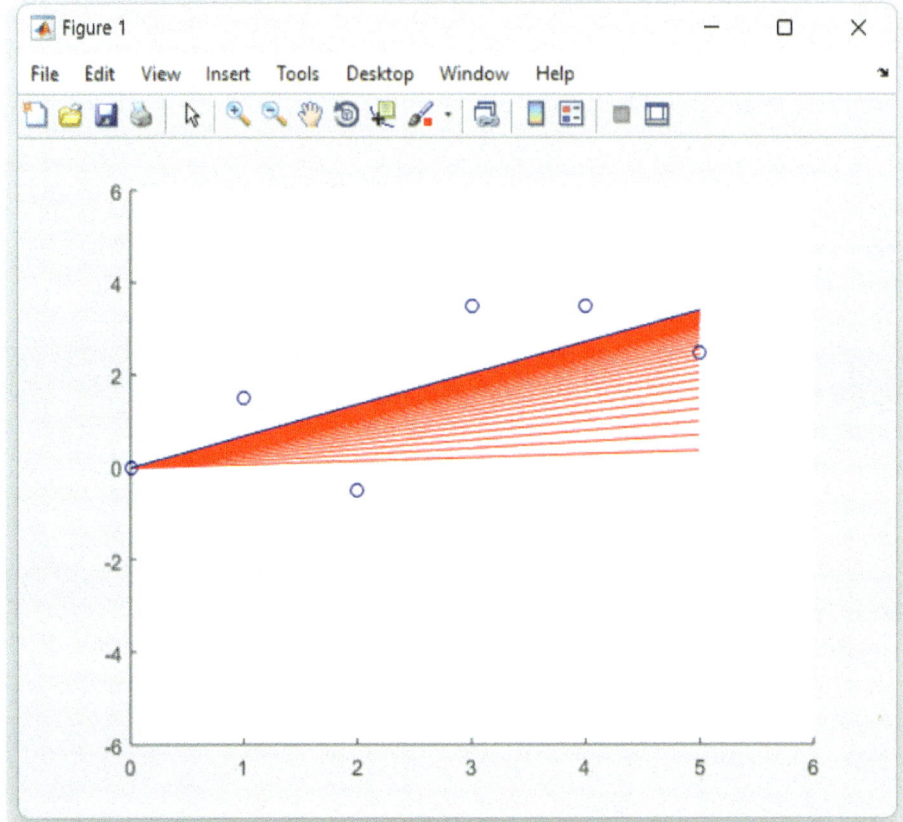

**Fig. 4.3** Graph showing the explored points and result of curve fitting using squared error

Additionally, it has been observed that using squared error provides a better algorithm performance, noting that Code 4.4 uses 100 iterations (niter), while with absolute error, we preferred to use 200. As an exercise, try testing the algorithm with squared error for the data in Table 4.3.

A detail worth highlighting from Code 4.4 is that within the iteration loop, the error is calculated as follows:

```
sum_d = sum( (y - m *x).^2 );
```

**Table 4.3** Alternative coordinates of the problem for a beeline

| x | 0 | 1 | 2 | 3 | 4 | 5 |
|---|---|---|---|---|---|---|
| Y | 0 | 1 | 2 | 3 | 4 | 5 |

## 4.2 The Squared Error

The dot before the exponentiation sign indicates to the compiler that this operation of squaring is intended for each element of a vector, not for squaring a vector itself, which would be mathematically incorrect, as only scalars and square matrices can be squared.

The dot is also used for element-wise multiplication and division, indicating that operations should be performed on each element rather than as matrix operations. However, when multiplying (m*x) in that same line, the dot is not necessary since *m* is a scalar (a number) and not a vector (an array of numbers). Even though we could add the dot without causing an error, in this case (m.*x), it is not necessary.

We also mention that squared error is differentiable, which is an advantage, and we'll now address the reason. But first, we must remember that if signals have discontinuities, this can cause errors when applying the gradient method, as seen in the EOQ example. However, if the signals have a smooth behavior, it's possible to obtain an expression for the derivative of the squared error and use it instead of calculating the gradient through the incremental procedure (h).

The squared error between a straight line starting at the origin and a series of points can be expressed as (4.3):

$$E^2 = \sum_{i=1}^{n} (y_i - mx_i)^2. \tag{4.3}$$

Therefore, the partial derivative of the error with respect to *m* can be expressed as (4.4):

$$\frac{\partial E^2}{\partial m} = 2 \sum_{i=1}^{n} (y_i - mx_i)(-x_i). \tag{4.4}$$

Code 4.5 solves the problem using the gradient descent method, employing squared error and Eq. (4.4) to calculate the gradient. Note that it's not necessary to define an increment h, and the gradient calculation is simplified to the first line within the for-loop of the iterations.

```
% Code 4.5 - Curve fitting (straight line - squared error)
clear; clc; close;    % reset

x = [ 0 , 1    , 2    , 3    , 4    , 5   ];
y = [ 0 , 1.5  , -0.5 , 3.5  , 3.5  , 2.5 ];

InitialSlope   = 0 ;         % Initial slope
LearningRate = 0.001 ;       % Learning rate
NumIterations = 100 ;        % Number of iterations

% Plot the original data
scatter(x, y, 'b');
axis ([ 0 6 -6 6]) ;   % adjust zoom
hold on;

for i = 1:NumIterations  % Iteration loop

    grad = 2*sum((y - InitialSlope*x).*(-x));
    InitialSlope = InitialSlope - LearningRate * grad ;

    plot(x, InitialSlope*x, 'r');
    pause(0.1);

end

plot(x, InitialSlope*x, 'b');
Sol_m = InitialSlope
```

## 4.3 What if the Line Does not Start at the Origin

Let us consider the line doesn't need to cross the origin; this increments the dimension of the problem to two dimensions. We need to find the slope and the starting point now. We can use the squared error in this case (we can also use the absolute value, but the squared error may have two different partial derivatives).

The use of the squared error can be extended to straight lines that do not intersect the y-axis at $y = 0$ but at $y = b$. The squared error could be expressed as (4.5).

$$E^2 = \sum_{i=1}^{n} (y_i - (mx_i + b))^2. \tag{4.5}$$

Now, we would have two partial derivatives of the squared error, one with respect to $m$ and the other with respect to $b$, see (4.6) and (4.7).

$$\frac{\partial E^2}{\partial m} = 2 \sum_{i=1}^{n} (y_i - (mx_i + b))(-x_i). \tag{4.6}$$

## 4.3 What if the Line Does not Start at the Origin

$$\frac{\partial E^2}{\partial b} = 2\sum_{i=1}^{n}(y_i - (mx_i + b))(-1). \qquad (4.7)$$

With these expressions, we can write a program that performs the curve fitting with the straight line and considers that the curve does not need to pass through the origin. Code 4.6 accomplishes this task.

```
% Code 4.6 - Curve fitting (straight line - squared error)
clear; clc; close;   % reset

x = [ 0 , 1  ,  2  , 3  , 4  , 5  ] ;
y = [ 1 , 2 , 3 , 4 , 5 , 6];

Initial_slope = 0 ;      % Initial slope
Intercept = 0 ;
Learning_rate_m = 0.001 ;   % Learning rate for m
Learning_rate_b = 0.005 ;   % Learning rate for b
Iterations = 200 ;    % Number of iterations

% Plot original data
scatter(x, y, 'b');
axis ([ 0 6 -2 8]) ;  % adjust zoom
hold on ;

for i = 1:Iterations % Iteration loop

    Gradient_m = 2*sum((y - (Initial_slope*x+Intercept)).*(-x)) ;
    Gradient_b = 2*sum((y - (Initial_slope*x+Intercept)).*(-1)) ;

    Initial_slope = Initial_slope - Learning_rate_m * Gradient_m ;
    Intercept = Intercept - Learning_rate_b * Gradient_b ;

    plot(x, (Initial_slope.*x + Intercept), 'r');
    pause(0.1);

end

plot(x, (Initial_slope.*x + Intercept), 'b');
Solution_m = Initial_slope
Solution_b = Intercept
```

Note that now the straight line solution does not pass through the origin. As an exercise, try testing the algorithm with the quadratic error for the data in Table 4.4.

**Table 4.4** Alternate coordinates of the problem

| x | 0 | 1 | 2 | 3 | 4 | 5 |
|---|---|---|---|---|---|---|
| Y | 1 | 2 | 3 | 4 | 5 | 6 |

You will be able to verify that the program offers a good solution ($m = 0.99$, $b = 1.02$); still, intuitively, we can imagine that the optimal solution would be ($m = 1$, $b = 1$). This is due to the parameters we are programming, mainly because we limit the number of iterations to observe the optimizer's behavior, that is, to see the graphs.

The reader can imagine that if, instead of 200 iterations, we program 20,000, the execution would take a long time while we wait in front of the computer. We can make many iterations in a short time if we omit the graphs; for example, the following code is the same as 4.6. However, the graphs within the iteration cycle have been commented out (not graphed each iteration, only at the end).

```
% Code 4.7 - Curve fitting (straight line - quadratic error)
clear; clc; close;    % reset

x = [ 0 , 1 , 2 , 3 , 4 , 5 ] ;
y = [ 1 , 2 , 3 , 4 , 5 , 6 ];

m       = 0 ;         % Initial slope
b       = 0 ;
alpha_m = 0.001 ;     % Learning rate
alpha_b = 0.005 ;     % Learning rate
niter   = 2000 ;      % Number of iterations

% Plot original data
scatter(x, y, 'b');
axis ([ 0 6 -6 6]) ;  % adjust zoom
hold on ;

for i = 1:niter % Iteration cycle

    grad_m = 2*sum((y - (m*x+b)).*(-x)) ;
    grad_b = 2*sum((y - (m*x+b)).*(-1)) ;

    m = m - alpha_m * grad_m ;
    b = b - alpha_b * grad_b ;

end

plot(x, (m.*x + b), 'b');
Sol_m = m
Sol_b = b
```

We can notice that the code runs almost instantly despite performing ten times more iterations than before (`niter = 2000;`); we can also observe that the solution was more precise. We can also comment out only the line that pauses (`pause(0.1);`), but the code does not run as fast as in the case where we comment out both lines. However, it runs much faster than plotting and pausing.

Code 4.8 performs the same function, but instead of defining the gradient and evaluating it as a function, it is calculated numerically. What was defined as a function was the quadratic error.

## 4.4 What if the Equation is Quadratic (or a Larger-Order Polynomial)

```
% Code 4.8 - Curve fitting (straight line - quadratic error)
% Gradient estimated numerically
clear; clc; close;   % reset

y = [ 0 , 1.5 , -0.5 , 3.5 , 3.5 , 2.5 ] ;
x = [ 0 , 1   , 2    , 3   , 4   , 5   ] ;

E2 = @(m,b) (y - (m*x + b)).^2;

m       = 0 ;          % Initial slope
b       = 0 ;
alpha_m = 0.001 ;      % Learning rate
alpha_b = 0.005 ;      % Learning rate
niter   = 200 ;        % Number of iterations
h = 0.001 ;

% Plot original data
scatter(x, y, 'b');
axis ([ 0 6 -6 6]) ;   % adjust zoom
hold on ;

for i = 1:niter % Iteration cycle

    m = m - alpha_m * sum(E2(m+h, b) - E2(m, b))/h;
    b = b - alpha_b * sum(E2(m, b+h) - E2(m, b))/h;

    plot(x, (m.*x + b), 'r');
    pause(0.1);

end

plot(x, (m.*x + b), 'b');
Sol_m = m;
Sol_b = b;
```

## 4.4 What if the Equation is Quadratic (or a Larger-Order Polynomial)

The same principle can be used for larger-order polynomials. Let's now imagine that the function is not a straight line but a quadratic function of the type (4.8).

$$y = ax^2 + bx + c. \tag{4.8}$$

The quadratic error would be defined as (4.9).

$$E^2 = \sum_{i=1}^{n} \left(y_i - \left(a(x_i)^2 + bx_i + c\right)\right)^2. \tag{4.9}$$

Following the procedure used previously, code 4.9 solves the quadratic interpolation problem using the gradient descent method.

```matlab
% Code 4.9 - Curve fitting (quadratic curve - quadratic error)
clear; clc; close;    % reset

y1 = [ 0.1  0.2  1   3   6   10 ] ;
x  = [ 0    1    2   3   4   5  ] ;

E2 = @(a,b,c) (y1 - (a*x.^2 + b*x + c)).^2 ;

a = 0 ;   b = 0 ;   c = 0 ;   % initial parameters

alpha = 0.001 ;   % Learning rate
niter = 200 ;     % Number of iterations
h = 0.01 ;

y2 = a*x.^2 + b*x + c;

% Plot original data
scatter(x, y1, 'b');
axis ([ 0 5.5 0 11]) ;   % adjust zoom
hold on ;
plot(x, y2, 'r');

for i = 1:niter  % Iteration cycle

    a = a - alpha * sum(E2(a+h, b, c) - E2(a, b, c))/h ;
    b = b - alpha * sum(E2(a, b+h, c) - E2(a, b, c))/h ;
    c = c - alpha * sum(E2(a, b, c+h) - E2(a, b, c))/h ;

    y2 = a*x.^2 + b*x + c;
    hold on ;
    plot(x, y2, 'r');
    pause(0.025) ;

end

    hold on ;
    plot(x, y2, 'b');

Sol = [a, b, c]
```

Finally, we must mention that there is a wide variety of curve fitting methods [5–7], and it is even possible to generate a curve with zero error. The cost of making the error zero is that the function is usually not a straight line defined solely by its slope (m) and its intercept (b). In fact, to fit a curve that passes through n points with zero error, a polynomial of degree (*n-1*) is necessary. This method is called polynomial interpolation, a topic that is beyond the focus of this book. However, there is a function in Matlab that allows us to use this in an automatic way.

## 4.4 What if the Equation is Quadratic (or a Larger-Order Polynomial)

The function is called (polyfit(x,y,N)) and only requires the vector of the independent variable (x), the vector of the dependent variable (y), and the order of the polynomial you want to fit (N). Indeed, if the order is one, then the function provides a straight line (just like our code), similarly, we can ask for a function of second order, third order, etc. The higher the order of the polynomial, the lower the error, until having a zero error when the order (N) equals the number of points provided minus one. For an order higher, the error will always be zero, and actually, there will even be more than one polynomial that can fit the curve. However, those are mathematical details that are out of the scope of this book.

As mentioned in this section, a couple of Matlab codes are provided that will allow the reader to use the function (polyfit(x,y,N)) if interested, as it is a very powerful tool for some engineering courses.

Perhaps once mastering the techniques studied in this course, the reader could attempt to program a polynomial interpolation function (without using the Matlab function).

For now, code 4.10 performs the polynomial interpolation for the same points we have used (see Fig. 4.1).

```
% Code 4.10 - Curve fitting (polyfit function)
clear; clc; close;    % reset

x = [ 0 , 1   ,  2   , 3   , 4   , 5   ] ;
y = [ 0 , 1.5 , -0.5 , 3.5 , 3.5 , 2.5 ] ;

p = polyfit(x,y,5)

x2 = [ 0 : 0.1 : 5 ] ;

Y_fit = p(1)*x2.^5 + p(2)*x2.^4 + p(3)*x2.^3 ...
      + p(4)*x2.^2 + p(5)*x2 + p(6) ;

scatter(x, y, 'b') ; % Plot the points
axis ([ 0 6 -6 6]) ; % adjust zoom
hold on ;
plot(x2, Y_fit) ;     % Plot the obtained polynomial
```

Upon running the code, the graph displayed in Fig. 4.2 is shown. We notice that the polynomial function passes exactly through the provided points. The reader is invited to change some points and observe that the function (polyfit(x,y,N)) always finds the perfect fit (Fig. 4.4).

Let's look at some details of the code. After clearing the memory and screen in the first line, which we call "reset," the code initially defines the points of the location of the cities (Fig. 4.1, Table 4.1). It uses the polyfit function, which generates the coefficients of the vector that passes through all the points defined by the $x$ and $y$ vectors, that is, through the point (0, 0), (1, 1.5), (2, − 0.5), (3, 3.5), (4, 3.5), and (5, 2.5).

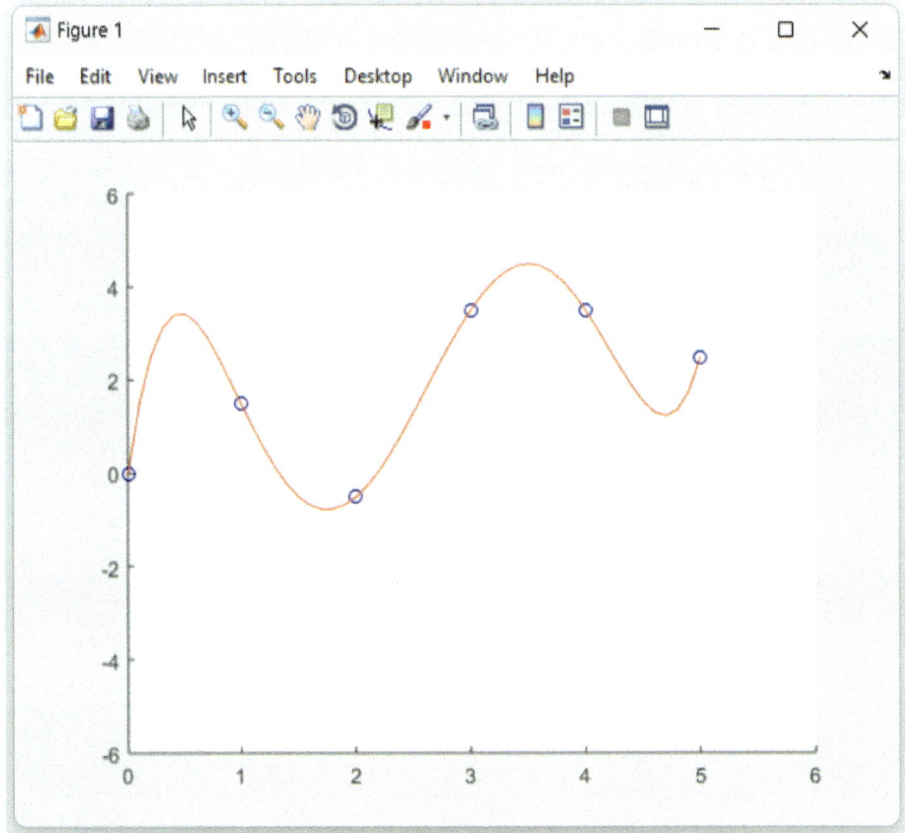

**Fig. 4.4** Graph of the curve fitting result with polynomial interpolation

```
% Code 4.10 - Curve fitting (polyfit function)
clear; clc; close;   % reset

x = [ 0 , 1   ,  2   ,  3   ,  4   ,  5   ] ;
y = [ 0 , 1.5 , -0.5 ,  3.5 ,  3.5 ,  2.5 ] ;

p = polyfit(x,y,5)
```

The coefficient vector p is used to generate a graph that can be overlaid with the city points, just as we overlaid the graphs of the line obtained by minimizing the quadratic error. It is possible to generate a 6-element vector with the defined $x$. However, it might be more interesting to see the graph in its full behavior, so we generate an $x$-axis with more points (51 points to be exact), starting with zero, and going from 0.1 to 0.1, up to 5.

```
x2 = [ 0: 0.1: 5 ];
```

The polynomial is generated with the following line:

```
Y_fit = p(1)*x2.^5 + p(2)*x2.^4 + p(3)*x2.^3 ...
      + p(4)*x2.^2 + p(5)*x2 + p(6) ;
```

Finally, we draw the points on the graph and plot the function obtained with the polynomial on top:

```
scatter(x, y, 'b')  ; % Plot the points
axis ([ 0 6 -6 6])  ; % adjust zoom
hold on ;
plot(x2, Y_fit) ;       % Plot the obtained polynomial
```

Additionally, Matlab has a function to plot the polynomial that does not require writing the polynomial in this way:

```
Y_fit = p(1)*x2.^5 + p(2)*x2.^4 + p(3)*x2.^3...
      + p(4)*x2.^2 + p(5)*x2 + p(6);
```

We are talking about the "polyval" function. The following code 4.11 performs the same function as 4.10, however, code 4.11 makes use of the "polyval" function.

```
% Code 4.11 - Curve fitting (polyfit function)
clear; clc; close;   % reset

x = [ 0 , 1    , 2    , 3   , 4   , 5   ] ;
y = [ 0 , 1.5  , -0.5 , 3.5 , 3.5 , 2.5 ] ;

p = polyfit(x,y,5) ;

x2 = [ 0 : 0.1 : 5 ] ;
y_fit = polyval(p, x2) ;

scatter(x, y, 'b'); % Plot the points
axis ([ 0 6 -6 6]); % adjust zoom
hold on;
plot(x2, y_fit);      % Plot the obtained polynomial
```

# References

1. Rao, S. S., Engineering optimization, theory and practice. Wiley, Fifth edition, 2019.
2. Bonnans, J. F., Gilbert, J. C., Lemaréchal, C., & Sagastizábal, C. A. Numerical optimization: theoretical and practical aspects. Springer Science & Business Media. 2006.

3. Zhu, T., Wang, S., Fan, Y., Hai, N., Huang, Q., Fernandez, C. (2024). An improved dung beetle optimizer- hybrid kernel least square support vector regression algorithm for state of health estimation of lithium-ion batteries based on variational model decomposition. Energy. 306.
4. Zeng, H., Wan, Ch., Zhong, W., Liu, T. (2025). Robust Integrative Analysis via Quantile Regression with Homogeneity and Sparsity. Journal of statistical planning and inference. 234.
5. Hamarashid, H.K., Hassan, B. A., Rashid, T.A. (2024).Modified-improved fitness dependent optimizer for complex and engineering problems. Knowledge-based systems. 300.
6. Hashim, F.A., Hussien. A.G. (2022). Snake Optimizer: A novel meta-heuristic optimization algorithm. Knowledge-based systems. 242.
7. Li. H., Tang. J., Pan. Q., Zhan. J., Lao. S.(2023).Ensemble of Population-Based Metaheuristic Algorithms. Computers, materials and continua. 76(3).

# Brief History and Classification of Metaheuristic Optimization Methods

## 5.1 Introduction

This chapter aims to summarize the way metaheuristic methods emerged and evolved, as well as the classification of different optimization methods according to their main characteristics. The concepts of heuristic and metaheuristic to solve everyday life problems expressed through mathematical models are described. Optimizing these problems represents the process of finding the *best solution* among a large set of possible solutions. At the end of this section, you will know the concepts of exploration and exploitation in the search space, as well as various strategies used to modify candidate solutions with the goal of approaching the best solution.

Process optimization is present in our daily lives. We are usually interested in minimizing or maximizing something. For example, we may be interested in minimizing our expenses and the distance we travel daily to work. Similarly, companies are interested in minimizing costs and maximizing profits, among other things.

In most of our activities, we are interested in finding *the best solution*. We use acquired knowledge and our experience to try to find solutions and choose the best among them. Similarly, various processes in nature develop in such a way that they always seem to achieve the best solution through a learned procedure that the elements participating in it have. For example, how does a group of monarch butterflies travel from North America to Michoacán, Mexico? It's as if there was a very established route and precise instructions to follow by the generations of those who undertook the journey. Similarly, the hunting that many animals do by stealthily approaching their prey and finding the best target, the cooling of certain materials until they achieve their best characteristics, and the exhaustive work of ant swarms or bees to build and create. That is, there are natural processes that seem to be simply perfect for their functioning in performing a task, so increasingly, humans are more interested in trying to decipher that perfection and in attempting to replicate it.

To discuss metaheuristic methods, it's important to define what a heuristic and metaheuristic are.

A Heuristic [1] is a strategy of instructions obtained through trial and error with the goal of finding optimal solutions for the problem under study in a reasonable time. Through the engineer's experience with the specific problem studied, these intuitive search rules are established.

A Metaheuristic [2] is a family of heuristics that come together to emulate the behavior of a process and thus achieve the best solutions in a reasonable time.

The idea is that an efficient and intuitive algorithm works within a reasonable time frame to find good solutions. With this type of algorithm, we cannot be sure of having a global solution for the search space. However, upon applying them, you will realize that they yield very good solutions in a short processing time.

The reader may ask why those methods are popular if they cannot guarantee to find the exact solution; well, there are problems in which it is impossible to determine what is the exact solution since the number of possible solutions is infinite, and we cannot try them all. There are other problems in which finding the exact solutions is so difficult and time-consuming that we prefer an algorithm that provides a good solution in a short time.

One of the first times the concept of heuristic was used was with Alan Turing and Gordon Welchman during World War II, when they designed an electromechanical device called the Bombe to aid the Enigma machine. With it, it was possible to decipher Enigma transmissions. The Bombe used a heuristic algorithm, so named by Turing, to search through various combinations and find the correct encoding of an Enigma message [3].

To be able to classify metaheuristic methods, it is important to know that many of the processes of interest for optimization are often represented through nonlinear and multidimensional mathematical functions. For this reason, it is of utmost importance to generate efficient strategies for solving these mathematical functions.

As mentioned in previous chapters, classical optimization methods are those based on the use of the derivative to achieve the algorithm's convergence to an optimal solution. These methods are quick for this type of optimization [4]. They are simple algorithms in their application; however, their use can be very particular due to the characteristics that the function to be optimized must fulfill. For the application of classical optimization methods, it is required that the function be differentiable and unimodal (a single minimum or maximum) [5]. Very simple situations can make our objective function nondifferentiable, or it could be a multimodal function (several maximums or minimums), making these methods not the best option to find solutions since the algorithm could get stuck in one of those local optima without the possibility of finding the global optimum.

Other tools emerged in previous decades as an alternative solution to the aforementioned. In the 1960s, one of the first methods appeared, basing its strategy on a completely stochastic process, the random search method [6]. The optimization process is carried out through the random generation of a single candidate solution, which evolves throughout the entire process. This process has been the basis for many metaheuristic methods. John Holland developed the genetic algorithm in 1960 [7]. This algorithm is based on the process of natural selection and Darwin's evolution. Holland was the first to establish operations of crossover, mutation, and selection of the best solution. These operators continue to be used to date in new algorithms (which will be addressed later). In the 1970s, works emerged reinforcing the idea of genetic algorithms. After the 1980s, metaheuristics experienced a significant boom.

The simulated annealing algorithm emerged by Kirkpatrick [8], inspired by the physical annealing process where a metal is heated to a temperature above its melting point and subsequently begins the cooling process until reaching a low temperature. In the same decade, the Tabu Search algorithm appeared [9]. In the 1990s, a technique inspired by swarm intelligence emerged, and to date, various metaheuristic algorithms have appeared, such as ant colony optimization [10] and particle swarm optimization, among others, that support solving complex mathematical problems.

Within the classification of optimization algorithms [11], we can find problems of single objective, multiobjective, unconstrained, constrained, unimodal, multimodal, linear, nonlinear, deterministic, and stochastic. In terms of metaheuristic algorithms, these are stochastic, based on a solution, or population-based (several solutions).

## 5.2 Optimization Methods Classification

To classify the various optimization techniques, let's recall how an optimization model is defined (Chap. 1). We will first think in two groups: first, the methods based on the derivative to achieve the search for the optimal solution, and second, the evolutionary methods. The latter does not use derivative information to perform the search strategy that will find the optimal solution.

The problem that needs to be minimized or maximized requires being expressed in terms of mathematical equations and constraints (if any) that calculate our response variable. The response variable could be the cost generated in the purchasing process or the waiting time in a process; this function is the one that will be minimized or maximized through a metaheuristic algorithm.

Minimize:

$$x \in \mathbb{R}^d \quad f(x_i), \forall x = 1, 2 \ldots d . \tag{5.1}$$

Subject to:

$$\delta(x_j), \forall x = 1, 2 \ldots J . \tag{5.2}$$

$$\omega(x_k), \forall x = 1, 2 \ldots K. \tag{5.3}$$

where Eqs. (5.1)–(5.3) are functions of a decision vector (Eq. 5.4), that is, the decision variables of which we wish to find their optimal values.

Equation (5.1) represents the objective function to optimize, and Eqs. (5.2) and (5.3) are the constraints to which the problem is subject.

$$x = (x_1, x_2, x_3, \ldots x_n). \tag{5.4}$$

Classical or evolutionary optimization techniques are used to optimize these mathematical models.

The classification presented in this chapter will be based on the techniques employed by the algorithms to optimize the mathematical models, depending on whether they use a single search agent or multiple agents. For example, within the classical methods, we can also find models of linear programming, nonlinear programming, stochastic programming, and dynamic programming; however, these models are not the focus of this book. For now, we will concentrate on the metaheuristic methods.

The search process to explore the space and find the best solution is of utmost importance in the algorithm. We can opt for methods where a single search agent is generated or by several agents where they will obtain information among themselves to follow the one that has the best solution.

The search space is the area over which the exploration will be carried out. For example, you may be interested in finding the deepest point in the sea; to date, the known deepest point is the Challenger Deep, which is almost 11,000 m deep, for reference, that is much more than the submerged Mount Everest (8848 m). However, this is the deepest known point in that exploration area, that is, in that search space. If humans had the capacity to fully explore the entire ocean with more advanced technology, possibly a new, deeper point could be found.

To explain the search space, let's take this example: the explored space at that time was in the northeast Pacific, including the Mariana Islands, near the Island of Guam; this Trench was determined as the search space (Fig. 5.2).

In this search area, a single agent or multiple agents can be generated. Let's think of the agent responsible for exploring this area of the Marianas. The first people who explored this search area were Jacques Piccard and Don Walsh in 1960 through the Trieste, which was a bathyscaphe (a submersible vessel for sea depths) (Fig. 5.3).

In this particular case, the exploration of the Mariana Trench can be classified as a single search agent method. Metaheuristic algorithms will support us in finding the solution to the mathematical model, in this case, illustrated with Eqs. (5.1)–(5.3). Through these algorithms (Fig. 5.1), initial solutions are applied to the model, thus generating a Population (P), that is, several search agents. The solutions are modified through the use of operators over a maximum assigned number of iterations (Total_Iter) until a feasible

## 5.3 Exploration and Exploitation

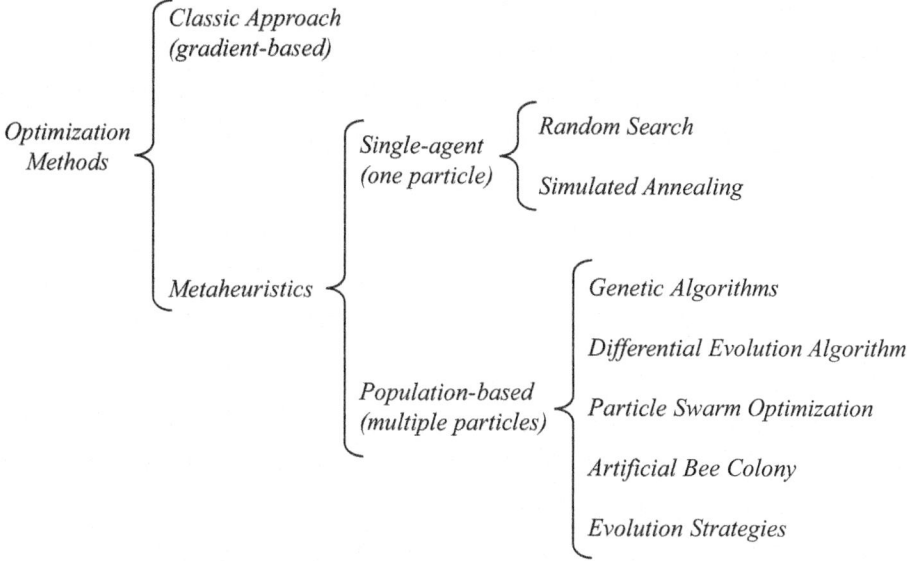

**Fig. 5.1** Classification of optimization methods

solution with excellent results is found. The following diagram represents the generic process for a population algorithm (Fig. 5.4).

## 5.3 Exploration and Exploitation

Exploration is defined as a search strategy for new solutions, which can be risky since the selected search area may be uncertain and, therefore, the quality of the solution to be found. With a good generation of the initial search agents and delineation of the search space, exploration can bring significant benefits to the quality of the solution. Exploitation refers to the use of solutions that have already been found through exploration to improve their solution quality. In exploration, we search for new solutions throughout our entire search space, while in exploitation, the best solutions found during exploration are locally refined. A good combination of both is necessary to explore good areas and then fine-tune (exploit) the best results [12].

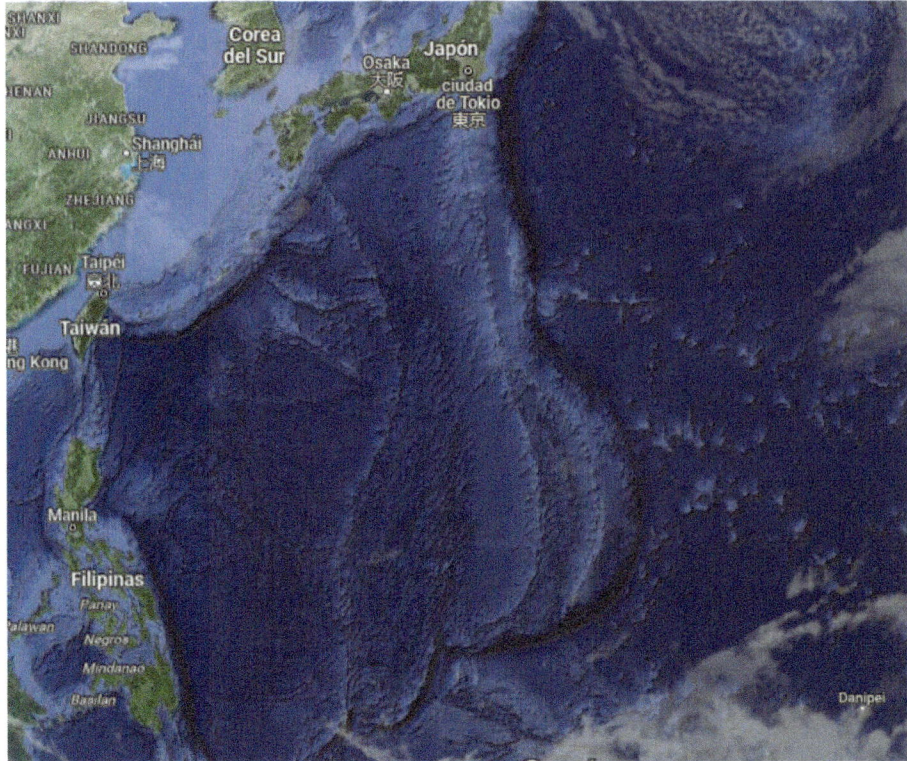

**Fig. 5.2** Search space

## 5.4 Basic Selection Techniques

Deciding whether a solution will be selected for the exploration of space and its exploitation is very important in the process. Probabilistic acceptance [13] is commonly used by various evolutionary methods. This operator refers to the ability to perform an action that is conditioned on a certain assigned probability. The goal is to select solutions so that search agents of higher quality have better chances of being chosen.

This operator is frequently used in various methods to keep the best search agents and thus refine the solutions. This operator can be understood as the act of performing action one if its value yields a valid probability. This probability is generated randomly under a uniform distribution U[0,1]; if the value of this random number, i.e., random probability, is better than the previously assigned probability as a requirement, then action 1 is executed.

Below is the Code 5.1 to make this selection.

## 5.4 Basic Selection Techniques

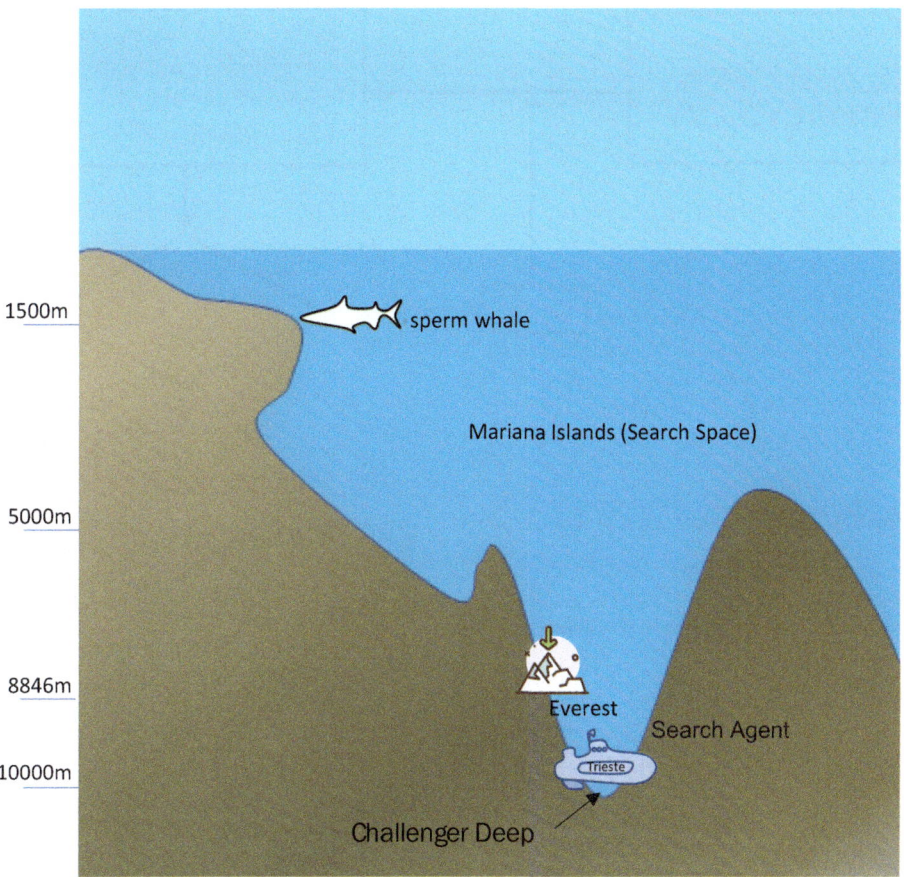

**Fig. 5.3** Agent in the search space

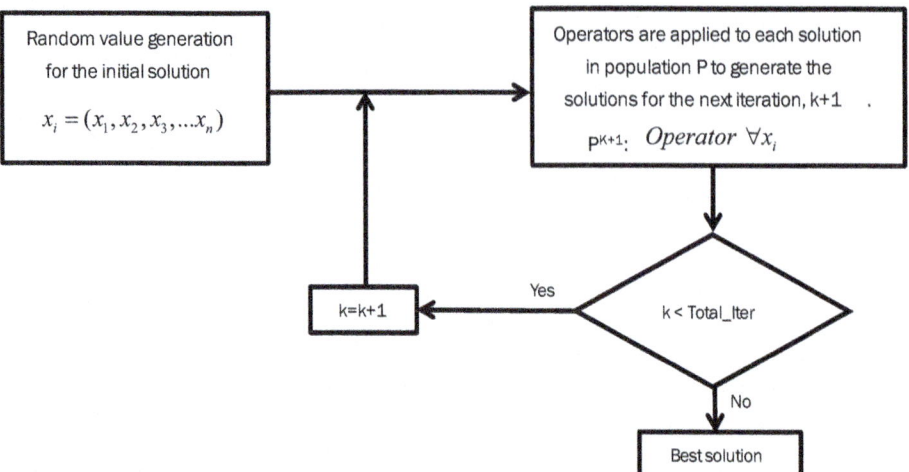

**Fig. 5.4** Generic optimization process in population algorithms

```
% Code 5.1 - Probabilistic Selection
clear ; clc ; close ;   % reset

task1=0;
notTask1=0;

% Iterative process to evaluate each agent
for i=1:100
    % Generate a random number between 0 and 1
    random= rand;
    % Probabilistic Acceptance Condition
    % the generated random number is compared against the requirement
    if (random<=0.8)
        % we will check how many elements meet the condition
        task1=task1+1;
    else
        % we will check how many elements do not meet the condition
        notTask1=notTask1+1;
    end
end
```

Once the code is executed, the selection of elements that meet the previously established condition can be made. The closer we set the requirement probability to 1, the more opportunities agents have to be selected. In the previous code, the results for those 100 agents were 79 times to select the task and 21 times not to execute it.

On the other hand, probabilistic selection is another operator frequently used by some metaheuristic algorithms. To explain this operator, it's necessary to introduce a new concept, the fitness of the solution. Fitness is the value obtained by the solution when it is evaluated in the objective function, so each agent will ultimately have fitness. Let's bring

## 5.4 Basic Selection Techniques

to this chapter the small model from the example in Chapter 1, where the goal was to minimize the cost for a bottled water seller.

$$(P) \quad \text{Min} \quad f(x) = \frac{50}{x} + \frac{0.05x}{2} + 5. \tag{5.5}$$

$$\text{s.t.} \quad 0 < x \leq 200. \tag{5.6}$$

Imagine that it wasn't possible to explore the entire search space, so we would have to generate random search agents, as has been explained in this chapter, and those variable values would be evaluated in the objective function (5.5). The following code shows how to generate some initial values for the first solutions, their evaluation, and therefore the obtaining of fitness.

```
% Code 5.2 - Fitness
clear; clc; close;   % reset

% Parameters
lowerBound=1;     % Lower limit for variable value
upperBound=200;   % Upper limit for variable value
x=[];

% Randomly generate 10 agents
for i=1:10
    x(i)= rand()*(upperBound-lowerBound)+lowerBound;
    f(i) = 50/x(i) + 0.05*x(i)/2 + 5 ; % Function
end
```

Table 5.1 shows ten values that were obtained in this run; you will get different values due to the randomness of the procedure.

**Table 5.1** Fitness values

| x      | Fitness |
|--------|---------|
| 37.78  | 7.27    |
| 181.07 | 9.80    |
| 195.97 | 10.15   |
| 88.34  | 7.77    |
| 23.11  | 7.74    |
| 52.35  | 7.26    |
| 82.34  | 7.67    |
| 119.38 | 8.40    |
| 53.18  | 7.27    |
| 120.97 | 8.44    |

These solutions are those to which different operators can be applied, with the aim of improving the solution quality. In metaheuristic methods, you must choose a solution xi from a population. Your choice should take into account the fitness of the solutions, with the aim that better-quality solutions have a better chance of being chosen. For this reason, it is important to know what the probability represented by each of the fitness values is.

The probabilistic selection operator aims for solutions with better fitness to be chosen. It should be noted that randomness (when contrasting against a random number) may allow solutions with not so good values to be selected. However, it is good for the algorithm's work to have good solutions (many of them) but also bad ones to allow continued exploration of certain areas.

Once we have the fitness of our solutions, we must consider what type of problem we are addressing. In this case, the presented model is a minimization model, so we will arrange the solutions from highest to lowest fitness, with the goal that, when calculating the cumulative probability, these solutions will have a higher chance of being chosen.

Table 5.2 shows the calculation of the probabilities, which must be calculated using the following formula:

$$P_{ij} = \frac{f(x_{ij})}{\sum_{i=1}^{N} f(x_{ij})}. \tag{5.7}$$

where, $f(x_{ij})$ represents the fitness of the solution $x_i$ in the family $\mathbf{x}(1, 2, 3 \ldots N)$ and in iteration $j$. It is important to remember that the algorithm will be executed on different occasions and that in each iteration, we have a family of solutions. Therefore, the probability for the solution $x_{ij}$ will be its fitness divided by the sum of all fitnesses.

Once the probability has been calculated, the cumulative probability must be considered. This represents the sum of the previous probabilities up to it. For example, the cumulative probability for the second solution is 0.244, which means 0.124 (previous

**Table 5.2** Probabilities

| x | Fitness | Probability | Cumulative probability |
|---|---|---|---|
| 37.78 | 7.27 | 0.124 | 0.124 |
| 181.07 | 9.80 | 0.120 | 0.244 |
| 195.97 | 10.15 | 0.103 | 0.347 |
| 88.34 | 7.77 | 0.103 | 0.450 |
| 23.11 | 7.74 | 0.095 | 0.545 |
| 52.35 | 7.26 | 0.095 | 0.640 |
| 82.34 | 7.67 | 0.094 | 0.733 |
| 119.38 | 8.40 | 0.089 | 0.822 |
| 53.18 | 7.27 | 0.089 | 0.911 |
| 120.97 | 8.44 | 0.089 | 1.000 |

probability) plus its probability of 0.120. The cumulative probability for the third solution is 0.347, meaning 0.124 plus 0.103 plus 0.120, and so forth.

This cumulative probability will be compared against a randomly assigned number; for this example, let's consider 0.6. If the cumulative probability is greater than this number, then the solution is chosen. Therefore, the best solutions will have a greater opportunity to be selected.

Just as there are these operators, there are others for the selection of good solutions, and thanks to randomness, it allows for the set of selected solutions to mostly contain good solutions but also some bad ones, with the goal of continuing to explore other sections of the search space.

## References

1. R. Colin. Heuristic Search Methods: A Review. Operational Research Society, Birmingham, UK, 1996, pp. 122–149.
2. K. Sörensen, M. Sevaux, and F. Glover, A history of metaheuristics, 2018. Handbook of Heuristics. Vol. 2–2, pp. 791–808. Book chapter
3. B. J. Copeland, Alan Turing's Automatic Computing Engine, Oxford University Press, 2005.
4. M. Pelikan, D. Goldberg,and F.G. Lobo.A survey of optimization by building and using probabilistic models. Computational Optimization and Applications. 2002, Vol 21(1), pp. 5–20.
5. P. Venkataraman, Applied Optimization with Matlab Programming, 2nd edition, 2009, John Wiley and Sons, Inc.
6. X.-S. Yang, Engineering Optimization, An introduction with metaheuristic application, John Wiley and Sons, Inc.
7. J. H. Holland "Outline for a logical theory of adaptative systems" J, ACM, vol 9, 1962
8. S. Kirkpatrick, C.D. Gellat, and M.P. Vecci. "Optimization by simulated annealing". Science. Vol. 220, Issue 4598. pp. 671–680. May 1983. ISSN 0036-807
9. F. Glover, and M. Laguna (1997) Tabu Search, Kluwer Academic Publishers, Boston.
10. M. Dorigo, and M. Birattari, Ant colony optimization, 2010, pp. 36–39, Springer US.
11. X. S, Yang. Engineering optimization: an introduction with metaheuristic applications. 2010. John Wiley & Sons.
12. T. Bäck & H.P.Schwefel. An Overview of Evolutionary Algorithms for Parameter Optimization. Evolutionary Computation. 1993, 1. 1–23.
13. E.V. Cuevas Jiménez, D.A. Oliva Navarro, M.A. Díaz Cortés, J.V. Osuna Enciso. Optimización: Algoritmos Programados con MATLAB. 2016, Colombia: Alpha Editorial.

# The EOQ Problem with Multiple Suppliers, Restrictions, and Volume Discounts

## 6.1 Introduction to the Inventory Administration Problem

In Chap. 1, John's problem was used to introduce some basic optimization concepts and a problem of particular interest to industrial engineering. John buys and sells gallons of water; he must manage his inventory to have gallons available to cover his demand, but he also learned to minimize the cost of managing his inventory. From an optimization viewpoint, the problem has been approached as a cost-minimization problem.

This chapter focuses on a more comprehensive version of John's problem. It is a problem that can be simple under certain considerations but can also be complex and has been widely studied in the last decades in industrial engineering, known as EOQ for its acronym in English (Economic Order Quantity) [1, 2].

The various variants of the EOQ problem [3] depend on the options considered, for example, taking into account that demand is constant or variable, that it can be calculated exactly (deterministically), or that we have an idea of the demand, but there is a random component (probabilistic). It is possible to consider whether there are volume discounts or not; discounts may be on the cost of products or transportation costs, and it is possible to have quality constraints for the parts or materials purchased. It can also be considered that there are various suppliers (or just one), and in the case that there are various suppliers, there is a possibility that they have (or do not have) production capacity constraints.

The number of variants of this problem, its particular form, which differs from other optimization problems that have been addressed, and its importance for industrial engineering make it convenient to dedicate a chapter to its study. In addition to studying this problem, these chapters will help us become familiar with some concepts from the application area.

John's case can be seen as a system with a flow of goods, in this case, gallons of water. Figure 6.1 shows the actors in this system with the product flow from left to right.

**Fig. 6.1** John's problem

John obtains the products from his supplier, whom he visits in a nearby town, through wholesale purchases (with a relatively large volume). John stores the product and sells it retail to his customers.

This simple system approximates the inventory management of a company; companies buy products, which can be basic raw materials in the case of primary manufacturers and can be made parts in the case of assemblers or final product manufacturers. The flow can also be of final products, in the case of distributors, wholesalers, and retailers. John is a retailer.

In all these cases, the optimization of inventory costs is carried out with the same concepts, although the mathematical expressions may vary in complexity. To generalize, instead of talking about John, we will talk about a company, see Fig. 6.2, where a decision-maker is located.

The company buys its products from its suppliers and can manufacture or sell the products as it buys them; whatever the process, it sells its products to its customers.

All the actors in the process can be companies, so to differentiate them, when we talk about the company or firm, we will refer only to the central actor (equivalent to John), where the decision-maker who will perform the optimization is located. The supplying companies will be simply called suppliers, and the companies or end-users will be called customers [4].

When we talk about purchases and optimization, a single product will be considered, although a large company normally requires many different products or parts. The optimization analysis is normally done separately for each of these.

**Fig. 6.2** The problem of a company

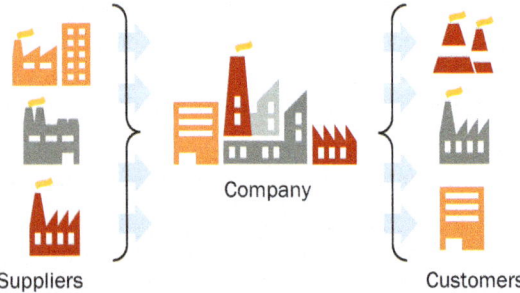

Let's look at some examples whose difficulty gradually increases, and we will also become familiar with the names of the variables.

## 6.2 Example 1. Two Suppliers Without Capacity Constraint

Let's assume that the company needs to buy 100,000 pieces a month of a product or part, which we will call demand $d = 100,000$; in John's case, his demand was five units (gallons of water) per day.

When we talk about thousands, we can use the prefix $K$, which means to multiply something by a thousand ($\times 10^3$), so $100,000 = 100 \times 10^3 = 100\,K$.

To meet the demand and to have those 100,000 pieces per month, we have two suppliers from which we can choose one, or we can even buy from both a combination of parts that add up to 100,000 pieces each month.

Supplier one offers the pieces at a cost of $p_1 = \$1.4$ per unit, while supplier two offers the units at a cost of $p_2 = \$1.6$ per unit.

If we had only this information, the choice would be simple: to buy from supplier one, as it has a cheaper cost per unit. However, we are also told that supplier one is more than double the distance from supplier two.

Transportation is handled by the company, which sends a truck to pick up the order. Due to the differences in distance, going by truck for an order from supplier one costs ten thousand dollars. We will call this cost $k_1 = \$10,000$, while in the case of supplier two, transporting an order costs \$4000, $k_2 = \$4000$. In both cases, we can assume that the company's truck is sufficient to transport an order of any size. Later on, we will take more complex considerations, but to start, this information is sufficient.

Besides the cost of the parts and the cost of transportation, there is a storage cost. Storing the parts in the company's warehouse has a calculated cost of $h = \$2$ per unit per month.

Now we ask ourselves, from which supplier should we buy? (it could be one or both). And similar to John's problem, how many pieces should we buy? It is assumed that each purchase will require one trip. Hence, making two purchases of 10,000 pieces from a certain supplier requires double the transportation cost than making one purchase of 20,000 pieces. However, as we saw with John's problem, making large purchases leads to higher storage expenses.

In this case, a limit on the suppliers' production capacity has not been set, so we can assume that either of them can provide 100,000 pieces per month. Thus, we could perform the analysis (similar to John's problem) for each of them and choose the one that results in a lower average cost.

Some of the equations we will use have been used in Chap. 1. However, we will write them all out for convenience. Furthermore, the explanation is different and with a broader focus. Later, but still in this chapter, these equations will be modified until reaching their final complexity.

To start, we will consider buying from supplier one and will analyze the optimal cost with this supplier. Subsequently, we will consider supplier two.

The objective is to determine the optimal order size ($Q$), but first, we usually try to mathematically express the cost as a function of $Q$ (which we will call $f(Q)$), to apply then some optimization technique that minimizes $f(Q)$ by finding the optimal value of $Q$.

To define $f(Q)$, we can consider that we have an unknown $Q$, which, although is unknown, knowing that the demand $d = 100{,}000$ pieces per month, we can assert that an order has to be made each period $T_C$, [5] calculated with (6.1),

$$T_C = \frac{Q}{d}. \tag{6.1}$$

For example, if we ultimately decide that the order size is $Q = 50{,}000$, then we will have to place an order every 0.5 months. The fact that the period has decimals does not represent a problem. We will assume that all orders made to a certain supplier will have the same size, as we can recall in John's problem, there is an order size $Q$, which minimizes the cost. So we will always use that optimal size, which at this moment we do not know (it is an unknown) for the stated problem, but we will determine it soon.

We can also define the average transportation cost as (6.2).

$$\text{Average transportation cost } i = \frac{k_i}{T_C} = \frac{d}{Q}k_i. \tag{6.2}$$

Remember that we have 2 transportation costs because the suppliers are at different distances, indicated by the letter $i$ in "$k_i$", so we will calculate the cost for each supplier separately, $k_1 = \$10{,}000$ and $k_2 = \$4{,}000$.

Equation (6.2) indicates that the smaller $Q$ is, the shorter the order cycle period ($T_C$) will be, necessitating more trips per month, increasing the transportation cost.

The storage cost is indifferent to the supplier because the storage is carried out at the company's facilities; as mentioned, it costs $h = \$2$ to store each part or unit for a month.

The storage cost can be analyzed with Fig. 6.3. This figure has some optimization considerations; for example, we assume that the inventory is filled with the order $Q$, once the parts have been depleted. As the units are being consumed, the warehouse level drops linearly until it reaches zero, when the inventory is replenished.

For example, if $Q = 50{,}000$, $T_C$ would be equal to 0.5 months, so twice a month, the inventory is replenished. When the inventory is filled, its maximum level is 50,000, and when the inventory runs out, it is refilled at the moment the inventory reaches zero. Intuitively, we might think this is risky; it would be safer to refill the inventory when it reaches, for example, 10,000. The maximum inventory level would be 60,000, and the minimum is 10,000. If an incident occurs that does not allow immediate restocking, there is a safety margin of 10,000 pieces.

**Fig. 6.3** Inventory Behavior

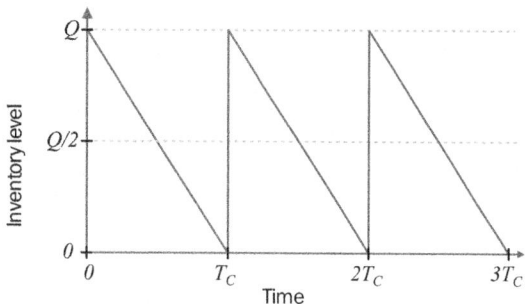

This would imply always having 10,000 more pieces than necessary. So, this safety margin would cost $20,000 a month. That's the cost of having 10,000 pieces stored all the time. Having this safety margin is a good idea; on the one hand, it will not be included within the optimization because this cost does not depend on the order number; in other words, it cannot be reduced with optimization. The only way to reduce it is to lower the storage cost per piece ($h = 2$), perhaps with some warehouse improvement strategy, or reduce the number of pieces kept as a safety margin.

Having a safety margin is a company decision, and although it can be analyzed deterministically, it is beyond the scope of this analysis. Thus, for analysis purposes, we will ignore the existence or not of a safety margin and will only account for the quantity of inventory that is being renewed.

Then, to calculate the storage or inventory cost [6], we need to calculate the average monthly quantity of pieces stored in the warehouse and multiply it by the cost of storing one piece per month. This average quantity of stored pieces can be expressed as (6.3).

$$\text{Quantity of pieces in inventory per period} = \frac{T_C Q}{2}. \qquad (6.3)$$

This is the number of pieces in the warehouse over a period $T_C$ (the area under the curve). If we multiply it by h and divide it by $T_C$, we will get the average monthly cost. This average inventory cost can be calculated as (6.4).

$$\text{Average inventory cost} = \frac{Q}{2}h. \qquad (6.4)$$

**Table 6.1** Summary of Parameters

|   | Definition | Example parameters |
|---|---|---|
| $r$ | Number of suppliers | 2 suppliers |
| $d$ | Demand | 100 thousand units a month |
| $h$ | Inventory cost | $2 USD per unit per month |
| $k_i$ | Transportation cost | $k_1 = 10{,}000$, $k_2 = 4000$, USD per order |
| $p_i$ | Cost of the parts | $p_1 = 1.4$, $p_2 = 1.6$, USD per unit |

Lastly, the cost of the pieces depends on the supplier. If we buy all the pieces from the same supplier, the average cost of the pieces can be calculated.

$$\text{Average cost of pieces} = p_i d. \tag{6.5}$$

The total average cost can be calculated by adding the three costs mentioned in (6.2), (6.4), and (6.5). The equation of $f(Q)$ is (6.6).

$$f(Q_i) = \frac{d}{Q_i}k_i + \frac{Q_i}{2}h + dp_i. \tag{6.6}$$

The subscript $i$ in Eq. (6.6) indicates that now there will be an optimal $Q$ for each of the two suppliers, so there will be an optimal $Q_1$ and an optimal $Q_2$.

This is the objective function $f(Q_i)$, and since we want to make it as small as possible, we are faced with a minimization problem. Table 6.1 shows a summary of the parameters in (6.6).

We could make a program to do the optimization, for example, using the gradient descent method, and run it twice, once with the parameters of supplier 1, and a second time with the parameters of supplier 2. Or we could make a single code that performs the optimization of both functions; in this case, we use one code for both functions. Code 6.1 minimizes the function (6.6) for both suppliers of the problem we have discussed. The code will be explained next.

## 6.2 Example 1. Two Suppliers Without Capacity Constraint

```
% Code 6.1 - Two Suppliers Without Capacity Restriction (Gradient)
clear ; clc ; close ;   % reset

x = [ 1 : 1 : 100001 ] ;    % generates x axis (q) for calculations

% Problem parameters
d = 100000 ;        % monthly demand
h = 2 ;             % storage cost per unit per month
k1 = 10000 ;        % transportation cost supplier 1
P1 = 1.4 ;          % cost of parts from supplier 1
k2 = 4500 ;         % transportation cost supplier 2
P2 = 1.6 ;          % cost of parts from supplier 2


CTr = k1.*d./x ;    % Average transportation cost
CAl = x.*h./2 ;     % Average storage cost
CPr = d*P1 ;        % Average product cost (gallons)
f1 = CTr + CAl + CPr ;   % Total cost for supplier 1
CTr = k2.*d./x ;    % Average transportation cost
CAl = x.*h./2 ;     % Average storage cost
CPr = d*P2 ;        % Average product cost (gallons)
f2 = CTr + CAl + CPr ;   % Total cost for supplier 2

plot(x, f1, x, f2, 'linewidth', 2) % plot the objective functions
hold on ;           % indicates not to replace the graph
axis ([ 0 100e3 1.8e5 2.8e5]) ; % adjust a convenient zoom

% Optimizer parameters
max_iter = 300 ;    % maximum number of iterations
q1 = 50000 ;        % initial point Q1
q2 = 50000 ;        % initial point Q2
alpha = 2000 ;      % learning rate
h = 1 ;             % delta of q

% Before starting, we draw the initial points
plot(q1, f1(q1), 'ro', 'MarkerSize', 4 ) ; % Supplier 1
plot(q2, f2(q2), 'ro', 'MarkerSize', 4 ) ; % Supplier 2

for i = 1:max_iter  % iterations

    grad1 = ( f1(round(q1+h)) - f1(round(q1)) )/( (q1+h) - q1 ) ;
    q1 = q1 - alpha * grad1 ;

    if q1<0         q1 = 0 ;        end % We avoid q going
    if q1>100000    q1 = 100000 ;   end % out of the allowed range

    grad2 = ( f2(round(q2+h)) - f2(round(q2)) )/( (q2+h) - q2 ) ;
    q2 = q2 - alpha * grad2 ;

    if q2<0         q2 = 0 ;        end % We avoid q going
    if q2>100000    q2 = 100000 ;   end % out of the allowed range

    plot(q1, f1(round(q1)), 'ro', 'MarkerSize', 4 ) ;
    plot(q2, f2(round(q2)), 'ro', 'MarkerSize', 4 ) ;
    pause(0.01);    % pause for 0.1 seconds to see the progress

end

format bank
Q1_opt = [ round(q1) f1(round(q1)) ]
Q2_opt = [ round(q2) f2(round(q2)) ]
```

If everything goes well, running the code will show Fig. 6.4 and the result in finding that, when using supplier 1, the optimal order size is 31,644, and the total average cost of maintaining the inventory is $203,243.57 per month. When using supplier 2, the optimal order size is 21,215, and the total average cost of maintaining the inventory is $202,426.41.

In other words, the optimal solution is to only buy from supplier 2, making orders of 21,213 units. Supplier 2 is the one that had the higher unit cost for the 100,000 pieces needed per month, which is a solution that does not seem intuitive.

It is easy to imagine that an intuitive solution like this would hardly occur to us. Intuitive solutions could be far from ideal. If the decision-maker had decided to order the 100,000 pieces from supplier 1 once a month, something that might seem intuitive, the monthly cost of maintaining the inventory would be $250,000. On the other hand, if someone decided to order 100,000 pieces a month from supplier 2, the total cost of maintaining the inventory would be $264,500.

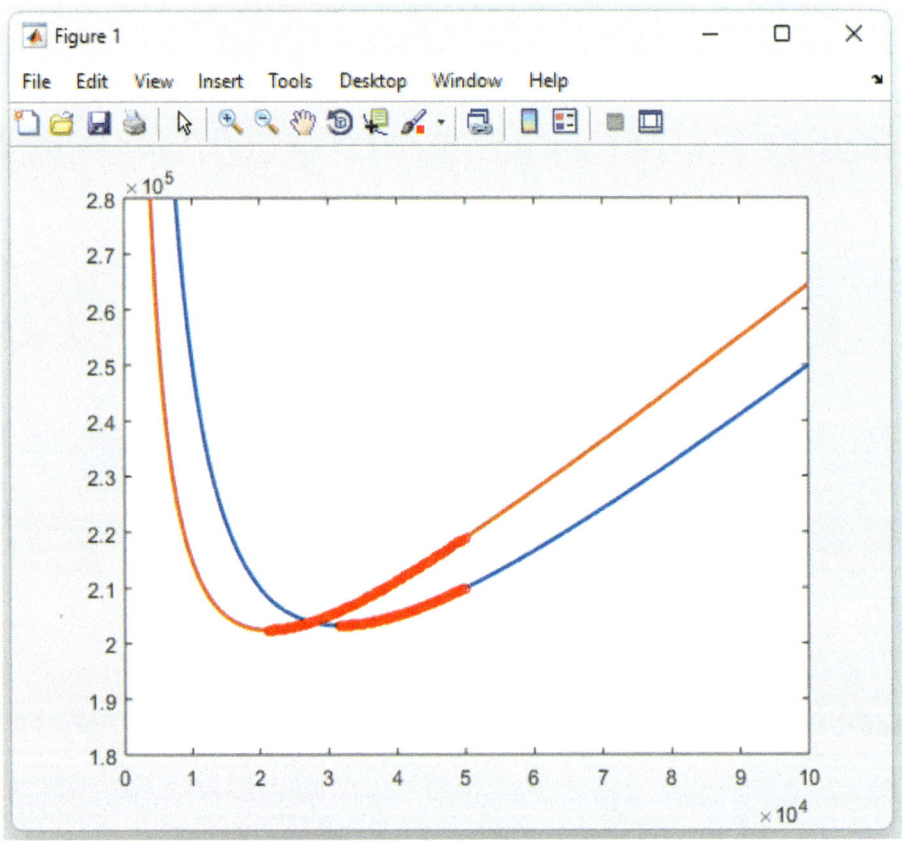

**Fig. 6.4** Graph showing the result of code 6.1

## 6.2 Example 1. Two Suppliers Without Capacity Constraint

Now, let us look at the code. The first lines reset the Matlab memory (deleting variables, clearing the screen, and closing open windows). In addition, they generate a vector of 100,001 values, which will be used as the *X-axis or Q-axis*. The reason for generating a vector one unit larger than the maximum number $Q$ can take is that in the gradient descent algorithm, the current $Q$ value $+$ 1 is used. This ensures that the algorithm cannot try to access a value out of range.

```
% Code 6.1 - Two Suppliers Without Capacity Restriction (Gradient)
clear ; clc ; close ;    % reset

x = [ 1 : 1 : 100001 ] ;     % generates x axis (q) for calculations
```

Subsequently, we introduce the system parameters according to the provided explanation and Table 6.1.

```
% Problem parameters
d = 100000 ;      % monthly demand
h = 2 ;           % storage cost per unit per month
k1 = 10000 ;      % transportation cost supplier 1
P1 = 1.4 ;        % cost of parts from supplier 1
k2 = 4500 ;       % transportation cost supplier 2
P2 = 1.6 ;        % cost of parts from supplier 2
```

After introducing the parameters, the objective functions are calculated. In this case, there are two because the cost is different for each supplier. In these lines, the costs are calculated according to Eq. (6.6).

```
CTr = k1.*d./x ;      % Average transportation cost
CAl = x.*h./2 ;       % Average storage cost
CPr = d*P1 ;          % Average product cost (gallons)
f1 = CTr + CAl + CPr ;    % Total cost for supplier 1

CTr = k2.*d./x ;      % Average transportation cost
CAl = x.*h./2 ;       % Average storage cost
CPr = d*P2 ;          % Average product cost (gallons)
f2 = CTr + CAl + CPr ;    % Total cost for supplier 2
```

Before starting with the optimization, we set the optimizer parameters.

```
% Optimizer parameters
max_iter = 300 ;      % maximum number of iterations
q1 = 50000 ;          % initial point Q1
q2 = 50000 ;          % initial point Q2
alpha = 400 ;         % learning rate
h = 1 ;               % delta of q
```

In the following lines, we enter the for loop for optimization, performing 200 iterations. It can be observed that both functions are optimized in the same for loop, but the optimization is done independently.

```
for i = 1:max_iter    % iterations

    grad1 = ( f1(round(q1+h)) - f1(round(q1)) )/( (q1+h) - q1 ) ;
    q1 = q1 - alpha * grad1 ;

    if q1<0           q1 = 0 ;         end % We avoid q going
    if q1>100000      q1 = 100000 ;    end % out of the allowed range

    grad2 = ( f2(round(q2+h)) - f2(round(q2)) )/( (q2+h) - q2 ) ;
    q2 = q2 - alpha * grad2 ;

    if q2<0           q2 = 0 ;         end % We avoid q going
    if q2>100000      q2 = 100000 ;    end % out of the allowed range

    plot(q1, f1(round(q1)), 'ro', 'MarkerSize', 4 ) ;
    plot(q2, f2(round(q2)), 'ro', 'MarkerSize', 4 ) ;
    pause(0.01);      % pause for 0.1 seconds to see the progress

end

format bank
Q1_opt = [ round(q1) f1(round(q1)) ]
Q2_opt = [ round(q2) f2(round(q2)) ]
```

We can notice that within the for loop, there are a couple of lines to limit q1 and q2, looking like this:

```
if q1<0           q1 = 0 ;         end % We avoid q going
if q1>100000      q1 = 100000 ;    end % out of the allowed range
```

These lines simply check if the order size does not exceed the limits to ensure that it is between 0 and 100,000 units, something that would cause an error in the program execution. These lines could be written in the following form, which is more similar to the normal way statements are written:

```
if q1<0
    q1 = 0 ;
end

if q1>100000
    q1 = 100000 ;
end
```

Writing the if statements in a single line helps to have a more compact code, although it does not impact the performance of the program.

## 6.3 Supplier Combination

Finally, before displaying the result, the line (format bank) asks Matlab to show the numbers without scientific notation, something optional, but this is the way we are accustomed to seeing the results of this type of problem.

## 6.3 Supplier Combination

We have analyzed an optimization case with 2 suppliers, and after optimizing the order size for each, we decided (with the help of a computer program and an optimization algorithm) that the optimal solution is to buy all the pieces from supplier 2 in orders of 21,215 units.

This solution turns out to be more economical than solutions that might seem intuitive, such as ordering 100,000 pieces every month (considering that we need 100,000 pieces per month). One question we didn't discuss is the idea of buying half the pieces from one supplier and half from the other or making a combination of both. In this chapter, we will analyze how to calculate the cost in cases where there is a combination of suppliers.

Besides satisfying curiosity, combining suppliers is necessary when there are capacity constraints, which is very common for large companies buying from smaller suppliers [7].

Imagine that the suppliers indicate that their capacity limit for producing parts is 60,000 pieces per month (either of them). Although supplier 2 is willing to sell us all the pieces of their production, we need supplier 1 because we would be short of 40,000 pieces per month to meet the demand. But how are the costs calculated? And how can an optimal decision be made? About the number of orders and the order size.

Let's start with the order cycle; if we now have pieces from two suppliers, Eq. (6.1) would evolve into Eq. (6.7).

$$T_C = \frac{\sum_{i=1}^{r} R_i}{d}. \tag{6.7}$$

where $R_i$ represents all the pieces purchased from supplier $i$, in our example, $r = 2$ (number of suppliers), so there will be an $R_1$, which are the pieces bought from supplier 1, and an $R_2$, which would be the pieces bought from supplier 2.

Suppose for illustrative purposes that we decide to make an order of 60,000 pieces from supplier 2, and an order of 40,000 pieces from supplier 1. This would mean that the order cycle would last $T_C = 1$ month, which is logical since the demand is 100,000 pieces, and we are buying a total of 100,000 pieces in those two orders.

But we must remember that the optimal order size might be a number that does not fit having an order per month. Suppose the optimal order size for supplier 1 turned out to be 30,000, and for supplier 2, it turned out to be 20,000. These numbers are similar to the result of the previous problem, but we will use round numbers as an illustrative example.

There's no problem in choosing $Q_1 = 30{,}000$ and $Q_2 = 20{,}000$; it's just that the order cycle would last 0.5 months (according to (6.7)), which is logical since $Q_1$ and $Q_2$ add up to half the monthly demand.

Now, suppose we want to order more pieces from supplier 2 because it turned out to be cheaper. In that case, we could make, in a single order cycle, 1 single order of pieces to supplier 1 and 2 orders to supplier 2, or a different combination, for example, 3 orders to supplier 2 and one order to supplier 1.

Then we will define the set of variables $J_1$, and $J_2$ as the number of orders made to supplier 1 and supplier 2, respectively. And we will continue using $Q_1$ and $Q_2$ as the order size of the orders made to supplier 1 and supplier 2, respectively.

In other words, if we decide that, in an order cycle, we will make 1 order of 30,000 pieces from supplier 1 and 2 orders of 20,000 pieces from supplier 2, our $J$'s and $Q$'s would be as follows: $J_1 = 1$, $J_2 = 2$, $Q_1 = 30{,}000$, $Q_2 = 20{,}000$. We can also use the vector form and think of both **J** and **Q** as vectors of size r (where r is the number of suppliers).

$$\mathbf{J} = \begin{bmatrix} J_1 \\ J_2 \end{bmatrix} = \begin{bmatrix} 1 \\ 2 \end{bmatrix}. \tag{6.8}$$

$$\mathbf{Q} = \begin{bmatrix} Q_1 \\ Q_2 \end{bmatrix} = \begin{bmatrix} 30000 \\ 20000 \end{bmatrix}. \tag{6.9}$$

Now we can define the variables $R_1$ and $R_2$, and the vector **R** that consists of both values, as (6.10).

$$\mathbf{R} = \begin{bmatrix} R_1 \\ R_2 \end{bmatrix} = \mathbf{J} \odot \mathbf{Q} = \begin{bmatrix} J_1 \\ J_2 \end{bmatrix} \odot \begin{bmatrix} Q_1 \\ Q_2 \end{bmatrix} = \begin{bmatrix} J_1 Q_1 \\ J_2 Q_2 \end{bmatrix}. \tag{6.10}$$

For our example, this vector **R** would be:

$$\mathbf{R} = \mathbf{J} \odot \mathbf{Q} = \begin{bmatrix} 1 \\ 2 \end{bmatrix} \odot \begin{bmatrix} 30000 \\ 20000 \end{bmatrix} = \begin{bmatrix} 30000 \\ 40000 \end{bmatrix}. \tag{6.11}$$

According to (6.7), the order cycle period would be 0.7 months for the example under study.

$$T_C = \frac{\sum_{i=1}^{r} R_i}{d} = \frac{R_1 + R_2}{d} = \frac{30000 + 40000}{100000} = 0.7. \tag{6.12}$$

It makes sense since the sum of units (6.11) is 70,000, and the monthly demand is 100,000.

## 6.3 Supplier Combination

After introducing Eq. (6.7), we mentioned that $R_i$ is the amount of pieces purchased from a supplier in an order cycle. At that time, we hadn't defined a formula for **R**, yet it's correct to say that $R_i$ is the quantity of pieces ordered from each supplier in a period of the order cycle. In the example under discussion, the order cycle period lasts 0.7 months, and in that period, 30,000 pieces are ordered from supplier 1, while 40,000 are ordered from supplier 2.

The transportation cost in an order cycle can be calculated as (6.13).

$$\text{Transportation cost in an order cycle} = \sum_{i=1}^{r} J_i k_i. \tag{6.13}$$

For the example we are discussing, this cost would be:

$$\sum_{i=1}^{r} J_i k_i = J_1 k_1 + J_2 k_2 = 1 \times 10000 + 2 \times 4500 = 19000. \tag{6.14}$$

In an order cycle of 0.7 months, we would make one trip costing $10,000 and two trips costing $4,500 each. But why are we interested in the average cost? To compare solutions with different order cycles. Since we have defined (6.7), the average transportation cost, which we will call $Z_1$, could be expressed as (6.15), which is (6.13) divided by the period of the order cycle.

$$Z_1 = \frac{1}{T_C} \left[ \sum_{i=1}^{r} J_i k_i \right] = \frac{d}{\sum_{i=1}^{r} R_i} \left[ \sum_{i=1}^{r} J_i k_i \right]. \tag{6.15}$$

The cost of the parts in an order cycle would be given by (6.16)

$$\text{Cost of the parts in an order cycle} = \sum_{i=1}^{r} R_i p_i. \tag{6.16}$$

It's simply the sum of each part multiplied by its cost. For our example, this would be:

$$\sum_{i=1}^{r} R_i p_i = R_1 p_1 + R_2 p_2 = 30000 \times 1.4 + 30000 \times 1.6. \tag{6.17}$$

$$\sum_{i=1}^{r} R_i p_i = 106000. \tag{6.18}$$

Remember, we are interested in the average cost to compare solutions with different order cycles, similar to (6.15), the average cost of the parts, which we will call $Z_2$, could be expressed as:

**Fig. 6.5** Inventory behavior in a TC period

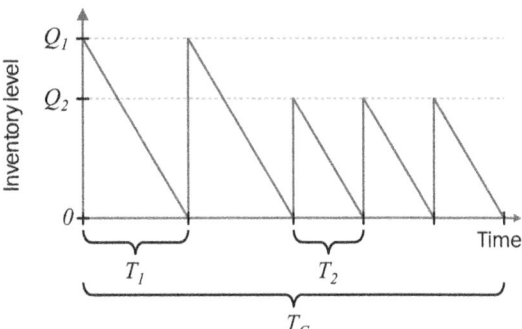

$$Z_2 = \frac{1}{T_C}\left[\sum_{i=1}^{r} R_i p_i\right] = \frac{d}{\sum_{i=1}^{r} R_i}\left[\sum_{i=1}^{r} R_i p_i\right]. \qquad (6.19)$$

The expression may seem complicated, just remember it is nothing more than mathematically expressing the concepts we have been discussing.

Now, let's examine the storage cost, which also varies. For the analysis, we will rely on the graph in Fig. 6.5. We need to calculate the average area under the curve, or the average number of pieces stored in the warehouse. In the previous case (Fig. 6.3), all orders were of the same size, but in this case, they may differ.

We will assume that orders are placed as the warehouse empties. If we made all orders together, the storage cost would be higher. We will also assume that we first complete orders from one supplier, and once we finish with the number of orders ($J_i$) from one supplier, we continue with the next, and so on. If we decided to alternate suppliers, we could observe that the area under the curve would not change.

In Fig. 6.5, there are two order sizes: two orders are made to one supplier and three to another. Thus, the time it takes for the warehouse to empty is different for each supplier's order. In the discussed problem, one order is made to supplier one and two orders to supplier two. The figure includes more orders to better appreciate the effect of combining suppliers.

The periods $T_1$ and $T_2$ could be calculated as follows:

$$T_1 = \frac{Q_1}{d}; \quad T_2 = \frac{Q_2}{d}. \qquad (6.20)$$

$$T_C = 2T_1 + 3T_2. \qquad (6.21)$$

The factors 2 and 3 are the number of orders made to each supplier, so (6.21) can be written in a general form as (6.22).

## 6.3 Supplier Combination

$$T_C = J_1 T_1 + J_2 T_2 = \sum_{i=1}^{r} J_i T_i = \sum_{i=1}^{r} \frac{J_i Q_i}{d} = \frac{1}{d} \sum_{i=1}^{r} J_i Q_i. \quad (6.22)$$

Equation (6.22) is equivalent to (6.12) and is another way to demonstrate the order cycle period.

The area under the curve (AUC) of Fig. 6.5 in a $T_C$ period would be expressed as:

$$\text{AUC} = 2\frac{Q_1 T_1}{2} + 3\frac{Q_2 T_2}{2} = 2\frac{Q_1}{2}\frac{Q_1}{d} + 3 \times \frac{Q_2}{2}\frac{Q_2}{d}. \quad (6.23)$$

We are summing the area of triangles. We have 2 large triangles and 3 small ones. The number of triangles for each supplier corresponds to the number of orders, so we could express (6.23) as (6.24).

$$\text{AUC} = J_1 \times \frac{Q_1^2}{2d} + J_2 \times \frac{Q_2^2}{2d} = \frac{1}{2d} \sum_{i=1}^{r} J_i Q_i^2. \quad (6.24)$$

Some texts use (6.24), considering the definition of $R_i = J_i Q_i$, which is an equivalent alternative form.

$$\text{AUC} = \frac{1}{2d} \sum_{i=1}^{r} J_i Q_i^2 = \frac{1}{2d} \sum_{i=1}^{r} \frac{R_i^2}{J_i}. \quad (6.25)$$

Finally, we can divide some of the expressions from (6.25) by the order cycle period ($T_C$) and multiply by $h$. We obtain the average storage cost of all pieces, which we will call $Z_3$.

$$Z_3 = \frac{1}{T_C}\left[\frac{h}{2d}\sum_{i=1}^{r}\frac{R_i^2}{J_i}\right] = \frac{d}{\sum_{i=1}^{r} R_i}\left[\frac{h}{2d}\sum_{i=1}^{r}\frac{R_i^2}{J_i}\right]. \quad (6.26)$$

Adding the average costs $Z_1$, $Z_2$, and $Z_3$, the average cost of inventory maintenance, which we will call $Z_T$, would be expressed as (6.27). This expression is the objective function we wish to minimize.

$$Z_T = \frac{d}{\sum_{i=1}^{r} R_i}\left[\sum_{i=1}^{r} J_i k_i + \sum_{i=1}^{r} R_i p_i + \frac{h}{2d}\sum_{i=1}^{r}\frac{R_i^2}{J_i}\right]. \quad (6.27)$$

Code 6.2 performs the calculations to obtain the average inventory maintenance cost with the described equations, which will be explained next.

```
% 6.2 - Average Cost Calculator for 2 Combined Suppliers
clear ; clc ; close ; format bank ; % memory reset

d = 100000 ;     % monthly demand
h = 2 ;          % storage cost per unit per month
r = 2 ;          % number of suppliers

Q = [ 31000 21000 ] ; % Solution to test
J = [ 1    3 ] ;

% Problem parameters
k = [ 10000 4500 ] ; % fixed transportation cost for each supplier
P = [   1.4 1.6 ] ; % cost of each piece by each supplier

R = J.*Q ;
frequency = d/sum(R) ;   % 1/frequency

Z1 = (frequency) * sum(J.*k) ;        % Average
Z2 = (frequency) * sum(R.*P);         % Average
Z3 = (frequency) * (h/(2*d)) * sum(R.^2 ./ J); % Inventory cost

Ztotal = Z1 + Z2 + Z3
```

Code 6.2 is relatively straightforward. The first lines perform a memory reset for Matlab, introduce the problem parameters, and remember that (format bank;) tells Matlab that, when displaying any result on the screen, not to use the scientific format, but rather the banking format, which we are accustomed to.

```
% 6.2 - Average Cost Calculator for 2 Combined Suppliers
clear ; clc ; close ; format bank ; % memory reset

d = 100000 ;     % monthly demand
h = 2 ;          % storage cost per unit per month
r = 2 ;          % number of suppliers
```

Subsequently, the solution to be tested is introduced. This is where we can modify if we want to test another solution, as well as the suppliers' parameters.

```
Q = [ 31000 21000 ] ; % Solution to test
J = [ 1    3 ] ;

% Problem parameters
k = [ 10000 4500 ] ; % fixed transportation cost for each supplier
P = [   1.4 1.6 ] ; % cost of each piece by each supplier
```

## 6.3 Supplier Combination

We now calculate the number of pieces purchased from each supplier (Eq. (6.19)), and the purchasing period. In fact, in this case, we can call it frequency because it is the inverse of the period. In the described equations, it is normally divided by the period, so if we calculate the inverse of the period, we can use that number as a factor that multiplies wherever it is necessary to divide by the period.

```
R = J.*Q ;
frequency = d/sum(R) ;   % 1/frequency
```

Note that **J** and **Q** are vectors, the dot before the asterisk indicates that the multiplication is done element by element (Hadamard product). In the second line, d is a scalar (equal to 100,000), but **R** is a vector. The instruction sum(R) performs the sum of all elements of the vector in parentheses, in this case, **R**. Although **J** and **R** are small vectors, each containing only two elements, the same lines would work for vectors with many elements.

Lastly, the average costs ($Z_1$, $Z_2$, and $Z_3$) are evaluated according to Eqs. (6.15), (6.19), and (6.26). Finally, they are added up to the total cost.

```
Z1 = (frequency) * sum(J.*k) ;       % Average
Z2 = (frequency) * sum(R.*P);        % Average
Z3 = (frequency) * (h/(2*d)) * sum(R.^2 ./ J); % Inventory cost

Ztotal = Z1 + Z2 + Z3
```

The semicolon (;) was not added to the last line because we want it to display that result in the command window.

We can use this code for solutions with a single supplier, but it would be necessary to change the vectors **Q**, **J**, **k**, and **P**, and use them as scalars.

With this code, we can now calculate the inventory maintenance cost when using combined suppliers. If we evaluate the idea of making an order to each supplier of $Q_1 = Q_2 = 50{,}000$ pieces, which would result in a $T_C = 1$ month. This would be equivalent to saying that we will evaluate the solution $\mathbf{J} = [1]$, and $\mathbf{Q} = [50{,}000, 50{,}000]$. The average cost (the objective function) would be equal to \$214,500 per month.

If we evaluate the solution mentioned at the beginning of this subsection $\mathbf{J} = [1, 2]$, and $\mathbf{Q} = [30,000, 20,000]$. The average cost (the objective function) would be equal to \$202,857.14 per month. This is not the optimal solution, though it is better than the previous ones. However, at this point, we have not mentioned how to obtain a solution; we have only analyzed how to calculate the costs when needing to buy a combination from different suppliers.

It's interesting to think about how we could obtain a solution, or even better, how we could obtain the optimal solution.

Let's recap a bit of what we have obtained. When suppliers have enough capacity to meet demand, it suffices to calculate the cheapest supplier, and that will be the best solution. However, sometimes suppliers do not have enough capacity to meet the demand, which is common in large companies that buy from small suppliers manufacturing the parts. When this occurs, it is necessary to use a combination of suppliers, and the total cost can be calculated with code 6.2.

## 6.4 Introducing Capacity Constrains

Before proceeding, it would be convenient to discuss the following. If we evaluate the solution $\mathbf{J} = [1, 3]$, and $\mathbf{Q} = [31,000, 21,000]$, the average cost (the objective function) would be \$202,702.13 per month. This cost is lower than other solutions with a combination of suppliers that have been mentioned; however, this solution requires that supplier 2 produces 67,021.28 pieces per month. We had mentioned imagining that the suppliers cannot produce more than 60,000 pieces per month.

In other words, when there are constraints of any kind, for example, capacity constraints [8], it's necessary to verify that the constraints are met. It's not enough to have a solution that results in a low cost using the cost calculator.

Code 6.3 calculates the average cost of maintaining the inventory, in addition, a capacity constraint is established, and the program evaluates if the introduced solution complies with the capacity constraint.

## 6.4 Introducing Capacity Constrains

```
% 6.3 - Average Cost Calculator for 2 Combined Suppliers
clear ; clc ; close ; format bank ; % memory reset

d = 100000 ;      % monthly demand
h = 2 ;           % storage cost per unit per month
r = 2 ;           % number of suppliers

Q = [ 31000 21000 ] ; % Solution to test
J = [ 1  2 ] ;

% Problem parameters
k  = [ 10000 4500 ] ; % fixed transportation cost for each supplier
P  = [   1.4 1.6 ] ; % cost of each piece by each supplier
Cp = [ 60000 60000 ] ; % Maximum monthly production capacity

R = J.*Q ;                % Calculate pieces produced per cycle
frequency = d/sum(R) ;    % 1/frequency, Calculate the cycle
Pp = (frequency)*R ;      % Calculate pieces produced per month

Z1 = (frequency) * sum(J.*k) ;    % Transportation cost
Z2 = (frequency) * sum(R.*P);     % Cost of the pieces
Z3 = (frequency) * (h/(2*d)) * sum(R.^2 ./ J); % Storage cost

Ztotal = Z1 + Z2 + Z3

for i = 1 : r
    if Pp(i) > Cp(i)
        disp(['Supplier ' num2str(i) ' exceeds Cp' ]);
    end
end
```

The code is very similar to the previous one, so we'll only comment on the differences. In the section where the problem parameters are set, the vector **Cp** is declared, containing the monthly production capacity of each supplier, in this case, 60,000 for each supplier is used as an example. Then the following lines appear.

```
R = J.*Q ;                % Calculate pieces produced per cycle
frequency = d/sum(R) ;    % 1/frequency, Calculate the cycle
Pp = (frequency)*R ;      % Calculate pieces produced per month
```

The first line calculates the vector **R**, which contains the number of pieces ordered from each supplier in an order cycle. However, the capacity of each supplier is given in pieces per month, so it's necessary to calculate how many pieces each supplier produces per month. The second line calculates the frequency of the order cycle (the inverse of the period), and finally, the third line calculates the vector **P**$_p$ (pieces produced), which contains how many pieces a supplier produces per month. If the order cycle period were one month, the frequency (one divided by the period) would also be one, but this code assumes that the order cycle period can be different from one month.

Lastly, after displaying the solution $Z_{total}$, a for loop appears that checks, element by element if **P**$_p$ is greater than **C**$_p$. If this occurs, it means that the solution being evaluated exceeds the production capacity of one of the suppliers.

```
for i = 1 : r
    if Pp(i) > Cp(i)
        disp(['Supplier ' num2str(i) ' exceeds Cp' ]);
    end
end
```

The evaluation of this constraint can be done in several ways. The advantage of the method used in code 6.3 is that it allows identifying which of the suppliers would face production capacity issues. We could refine the code to know the percentage of production capacity that the other suppliers have; readers are invited to try it.

In the standard form in which optimization problems are described, constraints usually come after the objective function, for example, as follows:

$$(\mathbf{P}) \quad \text{Min} \ Z_T = \frac{d}{\sum_{i=1}^{r} R_i} \left[ \sum_{i=1}^{r} J_i k_i + \sum_{i=1}^{r} R_i p_i + \frac{h}{2d} \sum_{i=1}^{r} \frac{R_i^2}{J_i} \right]. \tag{6.28}$$

$$\text{s.t.} \quad \frac{R_i}{T_C} \leq c_i, \quad \forall i = 1, \ldots, r, \tag{6.29}$$

This model is read as: The objective is to minimize the objective function described by Eq. (6.16), subject to (s.t.) the constraint described in Eq. (6.17).

## 6.5 Introducing Volume Discounts

In Chap. 1, John's problem became more complex when we introduced volume discounts. This made it impossible to use the gradient descent method, necessitating the use of a stochastic method, the random search or "Random Search".

As mentioned, it's natural for the unit price of goods to be lower when purchasing high volumes. This is what allows John's business, and many companies in the supply chain such as distributors or wholesalers and retailers, to be profitable. Companies prefer to have large volume sales. Therefore, volume discounts are a standard practice in all markets.

We will address an example where the company's suppliers offer volume discounts [9]. Additionally, we will consider that there are three suppliers, $r = 3$. The monthly demand remains $d = 100,000$ pieces per month, and the storage cost remains at $2 USD per piece stored per month.

The transportation costs are fixed in this example, meaning they do not depend on the volume of purchase. A later example will introduce the case where they do depend on the volume of purchase. In the previous example, the transportation costs were also fixed, but in this example, different values are used. The transportation cost $k_i$ for an order from each supplier $i$ is specified in Table 6.2.

We can see that supplier 1 has a higher cost for the transportation of each order. The most economical in terms of transportation is supplier 2, while supplier 3 has an intermediate cost.

The unit cost and discount ranges differ for each supplier. Supplier 1's costs are described in Table 6.3.

**Table 6.2** Summary of parameters

| $k_i$ | Supplier | Cost |
| --- | --- | --- |
| $k_1$ | 1 | $4000 USD per order |
| $k_2$ | 2 | $1000 USD per order |
| $k_3$ | 3 | $2500 USD per order |

**Table 6.3** Unit costs and discount ranges for supplier 1

| Pieces | Cost |
| --- | --- |
| 1 to 39,999 | $1.4 USD per unit |
| 40,000 to 54,999 | $1.31 USD per unit |
| 55,000 and above | $1.18 USD per unit |

**Table 6.4** Unit costs and discount ranges for supplier 2

| Pieces | Cost |
|---|---|
| 1 to 29,999 | $1.60 USD per unit |
| 30,000 to 49,999 | $1.46 USD per unit |
| 50,000 and above | $1.28 USD per unit |

**Table 6.5** Unit costs and discount ranges for supplier 3

| Pieces | Cost |
|---|---|
| 1 to 24,999 | $1.62 USD per unit |
| 25,000 to 44,999 | $1.53 USD per unit |
| 45,000 and above | $1.33 USD per unit |

From Table 6.3, we can observe that supplier 1 sells the pieces at $1.4 USD for low volumes. If someone buys more than 40,000 pieces, the cost decreases to $1.31 USD, and for volumes exceeding 55,000 pieces, the cost decreases to $1.18 USD per piece.

Similarly to Tables 6.3 and 6.4 shows the cost information and price ranges for supplier 2. And Table 6.5 shows the same information for supplier 3.

Code 6.4 calculates the average cost of inventory maintenance considering the 3 suppliers and the mentioned data.

## 6.5 Introducing Volume Discounts

```
% 6.4 - Average Cost Calculator for 3 Combined Suppliers
clear ; clc ; close ; format bank ; % memory reset

d = 100000 ;        % monthly demand
h = 2 ;             % storage cost per unit per month
r = 3 ;             % number of suppliers

%Q = [ 20000 30000 45000 ] ; % Solution to test
%J = [ 1 1 0 ] ;

Q  = [ 30000 20000 10000 ] ; % Solution to test
J  = [ 1 2 3 ] ;

% Problem parameters
k  = [ 4000    1000    2500 ] ; % transportation cost per order (fixed)
Cp = [ 60000 60000 60000 ] ; % Maximum monthly production capacity

P1(     1:  39999) = 1.40 ; % Supplier 1
P1( 40000:  54999) = 1.31 ;
P1( 55000: 100001) = 1.18 ;

P2(     1:  29999) = 1.60 ; % Supplier 2
P2( 30000:  49999) = 1.46 ;
P2( 50000: 100001) = 1.28 ;

P3(     1:  24999) = 1.62 ; % Supplier 3
P3( 25000:  44999) = 1.53 ;
P3( 45000: 100001) = 1.33 ;

R = J.*Q ;              % Calculate pieces produced per cycle
frequency = d/sum(R) ;  % 1/frequency, Calculate the cycle

Z1 = (frequency) * sum(J.*k) ;   % Transportation cost
% Z2 is the cost of the pieces
Z2 = (frequency) * ( R(1)*P1(Q(1)) + R(2)*P2(Q(2)) + R(3)*P3(Q(3)) );
Z3 = (frequency) * (h/(2*d)) * sum(R.*Q);  % Storage cost
Ztotal = (frequency) * (Z1 + Z2 + Z3 )

Pp = (frequency)*R      % Calculate pieces produced per month

for i = 1 : r
    if Pp(i) > Cp(i)
        disp(['Supplier ' num2str(i) ' exceeds Cp' ]);
    end
end
```

If everything goes well, we will notice that the average cost with the provided solution is $188,100 per month.

Note that the code includes at the end the part that warns if any of the suppliers exceeds the individual production capacity.

## 6.6 Introducing Quality Constrains

The EOQ problem is very broad, as mentioned at the beginning of this chapter, it has been studied for decades in industrial engineering. The problem can become larger and more complex as more data are added and as special situations that must be considered in the mathematical model are taken into account.

Now, it's time to analyze one of the considerations that can lead to more restrictions, or at least, to modify the mathematical model developed. This is the quality of the products sold by the suppliers. Normally, suppliers that offer a lower unit price have lower quality products than the more expensive suppliers. Quality considerations can be divided into two parts: long-term quality issues, such as the lifespan of the parts, and short-term quality issues, such as manufacturing defects that prevent the part from being used immediately.

Since the analysis focused on in this chapter is intended for the short-term production or sale of products, we will examine quality from an immediate perspective. The goal is to account for how much it affects that one supplier has more defective pieces than another.

For example, let's assume that the decision-maker is in a manufacturing company that requires parts to integrate them into a more complex product like a car; a certain part must have precise measurements to be used. If it does not have the specified measurements, it must be discarded. And the cost is borne by the company.

It's almost impossible to receive all the pieces perfectly. Despite quality controls, there are always some defective pieces. Thanks to advances in quality controls, the number of defective pieces tends to be a few pieces per million, and the acronym PPM is used to express this. In this book, the definition of a percentage quality index will be used, for example, 0.99, which means that 99% of the pieces meet the quality standards, this would be equivalent to having 10,000 defective PPM (10,000 defective parts per million).

In the mathematical development of the problem, there are various ways to introduce quality constraints. It is possible to introduce it in the form of a constraint, in a style similar to (6.17). A constraint that ensures quality compliance, another way is to think that those defective pieces would affect the calculation of the order cycle period (or its inverse, frequency).

We can consider that, to avoid a shortage, defective pieces must be accounted for in the order cycle. This idea was introduced by Alejo et al. [7].

In these cases, the following equation should be used to calculate the new order cycle.

$$T_C = \frac{\sum_{i=1}^{r} R_i q_i}{d}. \tag{6.30}$$

where $q_i$ is the quality index of supplier i, which could be, for example, 0.95, 0.99. This means that the order cycle is calculated based on the pieces that can be used (or sold) to meet demand.

## 6.6 Introducing Quality Constrains

**Table 6.6** Quality indices of suppliers

| $q_i$ | Supplier | Quality index |
|---|---|---|
| $q_1$ | 1 | 0.995 |
| $q_2$ | 2 | 0.990 |
| $q_3$ | 3 | 0.999 |

Recalling Eq. (6.12), repeated here as (6.31), suppose our demand $d = 100,000$ pieces, and we have two suppliers to whom we order 30,000 and 40,000 pieces per cycle. The order cycle period, called $T_{C1}$ in (6.31), would be 0.7 months.

$$T_{C1} = \frac{\sum_{i=1}^{r} R_i}{d} = \frac{R_1 + R_2}{d} = \frac{30000 + 40000}{100000} = 0.7. \tag{6.31}$$

Now suppose we want to consider quality, with supplier 1 having a quality factor $q_1 = 0.98$, and supplier 2 having a quality factor $q_2 = 0.99$. The order cycle that considers quality, called $T_{C2}$ in (6.32), would be equal to 0.69 months.

$$T_{C1} = \frac{\sum_{i=1}^{r} R_i q_i}{d} = \frac{R_1 q_1 + R_2 q_2}{d} = \frac{30000 \times 0.98 + 40000 \times 0.99}{100000} = 0.69. \tag{6.32}$$

The new order cycle will affect the total cost, as the average cost elements are divided by the order cycle period, or multiplied by the frequency, which is the inverse of the period.

Returning to the example of the 3 suppliers. Suppose in the example from code 6.4 that the suppliers have a quality index described by Table 6.6.

Code 4.4 needs only one line change to consider quality, which is changing the order cycle calculation from:

```
frec = d/sum(R) ;    % 1/frec, Calculamos el ciclo
```

To the following form, which is Eq. (6.30).

```
frec = d/sum(R.*q) ;    % 1/frec, Calculamos el ciclo
```

We notice that, when executed, the cost increases from $188,100 to $190,301.09 per month. The work cycle is reduced due to the percentage of pieces that must be discarded. And this increases the cost, the cost increase is implicitly calculated with the same average cost equations.

The cost increase is not significant, at least in percentage terms, but as mentioned at the beginning of the book, some companies have very high inventory management costs, and a small percentage can mean a lot of money. Additionally, in a competitive world, any advantage counts. Moreover, the more real considerations are taken into account, the more accurate our calculations will be.

So far, we've learned to mathematically model the EOQ problem, that is, John's problem, although with John's problem we took the simplest model, without a combination of suppliers. Now we've considered a combination of suppliers, which elevates the problem's dimension and modifies the equations, as the average cost must be calculated simultaneously considering all suppliers. Volume discounts were also introduced, adding local optima to the problem, complicating its solution with gradient-based algorithms. Lastly, quality considerations were introduced, in the form of a percentage of perfect pieces, and we observed how this modifies the average cost.

But we leave a mystery unsolved; we did not present a code that obtains the solution (we only modeled the cost). In the following chapters, we will use modern optimization techniques to find the solution to this intricate, interesting, and very important problem within industrial engineering, the EOQ problem.

## References

1. Schwarz, L. B. The economic order-quantity (EOQ) model. Building Intuition: Insights from Basic Operations Management Models and Principles, 2008, 135–154.
2. Chikaputri, K.K., Yudhistira, G.A., and Qurtubi. omparison Analysis of Economic Order Quantity (EOQ) Method and Min-Max Method on Inventory Management. AIP Conference Proceedings. 2023.
3. Raju, U. A review of Economic Order Quantity modelling, their extensions and applicability. Journal of Physics: Conference Series. 2024.
4. Alejo-Reyes, A., Mendoza, A. and Olivares-Benitez, E.Inventory replenishment decisions model for the supplier selection problem facing low perfect rate situations. Optimization Letters. 2021, 15-5, 1509-1535.
5. Mendoza, A., and Ventura, J.A. Modeling actual transportation costs in supplier selection and order quantity allocation decisions. Operational Research. 2013, 13-1, 5-25.
6. Mendoza, A., and Ventura, J.A. Estimating freight rates in inventory replenishment and supplier selection decisions. 2009, 1, 3-4, 185-196.
7. Alejo-Reyes, A., Olivares-Benitez, E., Mendoza, A. and Rodriguez, A.Inventory replenishment decision model for the supplier selection problem using metaheuristic algorithms. Mathematical Biosciences and Engineering. 2020, 17-3, 2016-2036.
8. Alejo-Reyes, A., Mendoza, A. and Olivares-Benitez, E.A heuristic method for the supplier selection and order quantity allocation problem. Applied Mathematical Modelling. 2021, 90, 1130–1142.
9. Alejo-Reyes, A., Mendoza, A., Cuevas, E., Alcaraz-Rivera, M. A Mathematical Model for an Inventory Management and Order Quantity Allocation Problem with Nonlinear Quantity Discounts and Nonlinear Price-Dependent Demand. Axioms. 2023, 12.6, 547.

# 7 Probability Distributions and the Random Search Method

## 7.1 Introduction

In Chap. 1, we briefly introduced the Random Search method. John's problem became complicated when we introduced the concept of volume discounts, and it was not possible to solve it with the gradient descent method. So, we used Random Search and were able to obtain a good result with code 1.10.

In code 1.10, we used the (rand) function and explained in a simple and brief manner that this function randomly returns a number between 0 and 1. This information is correct, although it might be convenient to have more information on this, for example, that the (rand) function provides a uniform probability distribution [1]. It would be good now to explain what that means.

Additionally, there is another function in Matlab, which is most used for probabilistic algorithms, namely the (randn) function (with an 'n' at the end). The (randn) function offers a normal probability distribution [2], also known as the Gaussian distribution or the bell curve.

This chapter is intended to briefly explain these concepts. And it presents some variants of the Random Search algorithm, which are very useful.

The variants of the Random Search algorithm [3] that will be addressed are Local Random Search and Adaptive Random Search. The first variant, instead of searching randomly but with a uniform distribution across the entire solution range, uses a normal distribution (bell curve) focusing on a certain search region (with higher probability). The second variant uses the same principle, but gradually reduces the probable search range (the standard deviation) to focus gradually on a certain region.

These algorithms will be tested with some of the problems addressed previously, for example, the problem of minimizing the Bohachevsky function [4].

## 7.2 Probability Distributions

Reading this chapter will help us understand probability distributions [5] and the nature of the (randn) function.

A probability distribution is a graph that helps us know how likely it is to get a certain result in a certain event. Suppose we have a normal die with 6 faces, numbered from 1 to 6. If we throw the die, it can land on any of the 6 faces, and it's equally likely to land on face 1, compared to face 2, face 3, etc. All the faces have the same probability of ending up on top.

Figure 7.1 is a probability distribution graph showing the distribution of probabilities for the outcome of tossing a die into the air. What Fig. 7.1 indicates is that the result can be one of 6 options, 1, 2, 3, 4, 5, or 6. And each of the results has the same probability of occurring, which is (1/6). A distribution of this type, in which all the options in the range have the same probability of occurring, is called a *uniform distribution*.

The (rand) function in Matlab returns a number between 0 and 1, for example, 0.23, 0.55, etc. It has a uniform distribution, meaning there is no number more likely to occur than another. Strictly speaking, we must mention that this function can generate numbers very close to 0 (for example, 0.0001) and to 1 (for example, 0.9999), but not exactly 0 or 1, as it is not programmed for such. Anyway, for practical purposes, we should expect a number between 0 and 1.

But what happens if we don't roll one die but two, and we establish that the result is the sum of what both dice indicate? This is the most common way to use a pair of dice. We know the result can be one of 11 outcomes: 2, 3, 4, 5, 6, 7, 8, 9, 10, 11, or 12.

Figure 7.2 shows us a graph of the possible results of this experiment.

On the left side of Fig. 7.2 it says Die 1, and it shows the 6 possibilities of the result of die 1. These are ordered in the rows (horizontal). Below are those of die 2, in the columns (vertical), and in the center box, there are the sums of the results, at the intersection of the rows with columns. There are 36 possible outcomes of this experiment, or combinations of the dice. However, these 36 possibilities contain the 11 numbers we mentioned (from 2 to 12). This means some of the results repeat, for example, having 2 on die 1 and 3 on die 2 is the same as having 3 on die 1 and 2 on die 2.

The chart in Fig. 7.2 helps us visualize the most repeated results; some boxes are marked with gray just to discuss them and distinguish them easily. Let's first look at the result in the top left, which is marked with light gray, the result of 2. It's unlikely to get 2

**Fig. 7.1** Probability distribution of the result of rolling a die

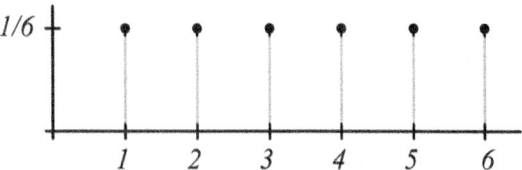

## 7.2 Probability Distributions

**Fig. 7.2** Possible outcomes of rolling two dice and adding the result

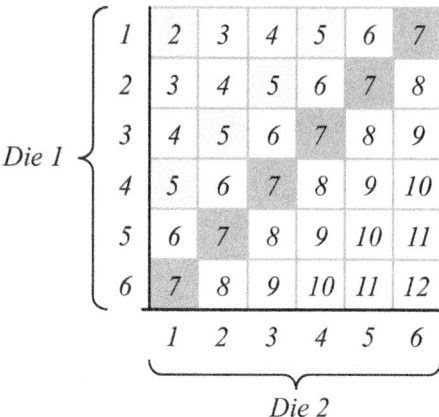

as a result because both dice have to land on 1, exactly, and there is no other option. From experience in board games, we can assert it's an uncommon result, the same happens to 12; it's an uncommon result.

But the explanation lies in probabilities. Of the 36 possible outcomes, only one results in 2, its probability is 1/36 (we can say 1 out of 36). The same happens to the number 12, it's unlikely because only one combination of the 36 results in 12, its probability is also 1/36.

Compared to 2 or 12, it's more likely to get, for example, a 5 because out of the 36 possibilities, 4 of them result in 5 (also marked in light gray in Fig. 7.2), that is, if die one shows 4, and die two shows 1, or if die one shows 3, while die two shows 2, the probability of getting a 5 is 4/36 (four out of thirty-six), 4 times more likely than getting a 2 or 12.

Seven is the most probable outcome, with 6 combinations of the dice resulting in 7, so it has the best probability of all outcomes, at 6/36.

Figure 7.3 shows the probability distribution of the outcome of our experiment of rolling 2 dice and adding their results. This distribution reflects what we have observed and discussed from Fig. 7.2.

**Fig. 7.3** Probability distribution of the outcome of rolling two dice

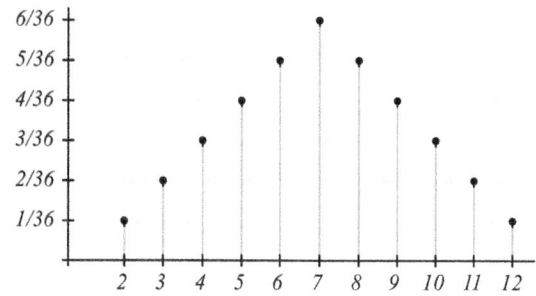

When the experiment has few options, like rolling two dice, which has 36 possibilities, it might be simple to state the information in words, but in options with more possibilities, the probability distribution graph is a very powerful tool. From Fig. 7.3, we can easily see that 7 is the best option, with a 6/36 probability. While 1 and 12 are the worst, 6 and 8 are also good options with 5/36.

Two important points from this discussion are (i) if you have to bet on a value in the two dice experiment, it's most advisable to bet on 7; 6 and 8 are also good options, the worst options are 2 and 12, as they have the lowest probability of appearing. (ii) The second point is that in the field of probabilities, there are these graphs, the probability distributions, like Figs. 7.1 and 7.2, that help us visualize this type of information.

The generic name for this type of graphs (like Figs. 7.1 and 7.3) is a histogram. A histogram [6] is a way of showing data or information in a graph, and it is used especially when we want to see how many times a value is repeated, or how the values of the set are distributed.

Histograms can be generated not only for a repeating value but also in intervals or classes, for example, people's heights can be grouped by exact heights in centimeters, or in groups, for example, with intervals of 5 cm or 10 cm.

There is a particularly important distribution or histogram for the topic of probabilities, and for probabilistic optimization algorithms, the normal probability distribution, also known as the Gaussian distribution or bell curve. It is characterized by having a symmetric bell shape and is fundamental in the theory of statistics due to its wide application in a variety of contexts, from physics to economics and social science.

The normal distribution is present in nature, in data such as people's height and weight. The characteristics of a given natural population tend to be distributed in a Gaussian bell curve. Figure 7.4 shows the shape of this distribution with its most common parameters, centered at zero, with a peak of 0.4.

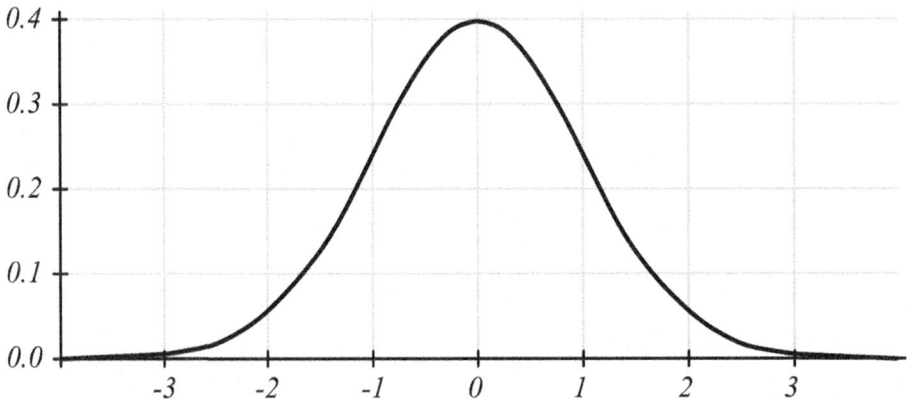

**Fig. 7.4** Normal distribution or Gaussian bell of the (randn) function

## 7.2 Probability Distributions

This function interests us because in the algorithms we use, we will normally use this Matlab function (randn), which is different from the one we used before (rand).

The study of probabilities is a very interesting and extensive area. Although it is not the objective of this chapter, the main goal is to understand what probability distributions are, and to know that when we use the (randn) function, the result will be a random number, but we can have a clear idea of its possible outcomes if we observe the probability distribution in Fig. 7.4.

Let's pause to review and remember the direction we are taking. We analyzed John's problem without volume discounts, and we solved it with the gradient method. We introduced volume discounts, where the gradient method encountered difficulties, and we migrated to a probabilistic method, the random search method.

Metaheuristic optimization algorithms can be seen as the right option for problems where the gradient method cannot find the result. Code 1.10, which solves John's problem with the Random Search method, makes use of the (rand) function, which throws a number between 0 and 1 with a uniform probability distribution, see Fig. 7.5. What we can see in this distribution is that the (rand) function will give us a value between 0 and 1. It's a continuous distribution, meaning it can give numbers with many decimals, like 0.1, or 0.123, 0.755, etc. And the probability of giving us 0.1 is the same as giving us 0.343, or 0.978, etc. In other words, it will give us a value between 0 and 1 chosen at random, and no number has a higher probability than another.

The most used function in probabilistic algorithms is not the (rand) function, but the (randn) function, which is different. The (randn) function has a normal probability distribution, similar to Fig. 7.4. This means it can give us a positive or negative value, which may be between $-1$ and 1, but it also has values outside this range, for example, it can give $-1.5$, $-2.1$, but this is less likely.

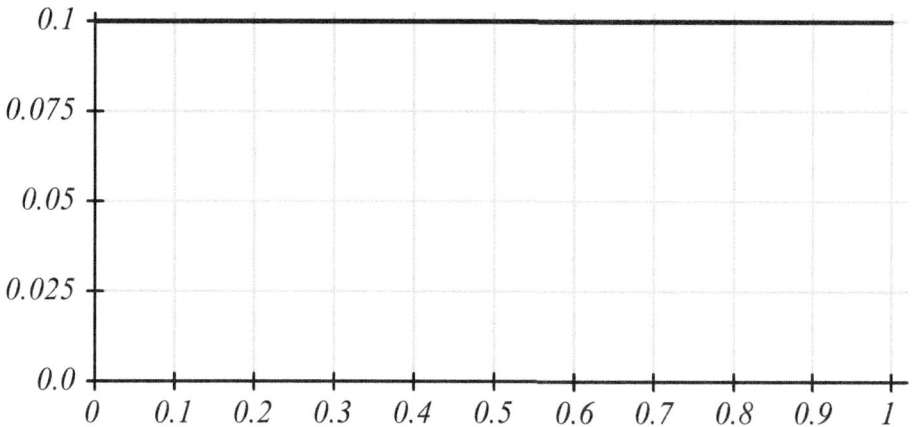

**Fig. 7.5** Uniform probability distribution of the (rand) function

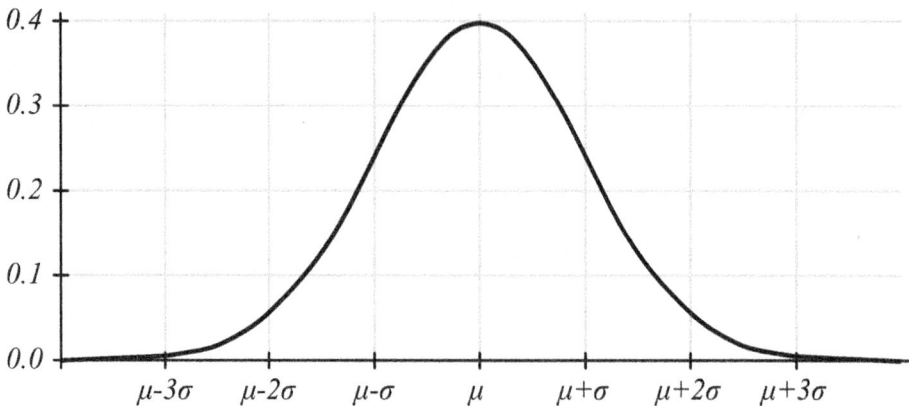

**Fig. 7.6** Normal distribution with parameters (randn)

In other words, the (randn) function will give us a random value, but not all values have the same probability. The most likely is that it will be close to zero (the probability distribution is not uniform).

Moreover, there are 2 factors we can use to adjust the Gaussian bell. Remember that in Code 2.7 (in Chap. 2), we wanted to choose a value between 1 and 200, instead of a value between 0 and 1, so we used the (rand) function multiplied by 199 and added 1. Similarly, we can adjust the (randn) function to take values in a different range than in Fig. 7.4. In fact, strictly speaking, the normal distribution has the form described in Fig. 7.6, where the central parameter $\mu$ (mu) is called the mean, and the parameter $\sigma$ (sigma) is called the standard deviation.

The mathematical equation that describes the normal distribution is given in formula (7.1).

$$f(x) = \frac{1}{\sigma\sqrt{2\pi}} e^{-\frac{(x-\mu)^2}{2\sigma^2}}. \tag{7.1}$$

where, as mentioned, $\mu$ is the mean, $\sigma$ is the standard deviation, and the constant parameters are $\pi$ (3.14159...) and e (2.71828).

The most common is for the mean ($\mu$) to be zero and the standard deviation ($\sigma$) to be one; this is what we refer to in Fig. 7.4, with the most common parameters. With these parameters, the probability that the response from (randn) is between $-1$ and 1 is 68%, the probability that the response is between $-2$ and 2 is 95%, and the probability of being between $-3$ and 3 is 99.7%.

The following lines of a Matlab code plot the function described in formula (7.1), evaluating the function directly (this is not a probability exercise, just an evaluation of (7.1)).

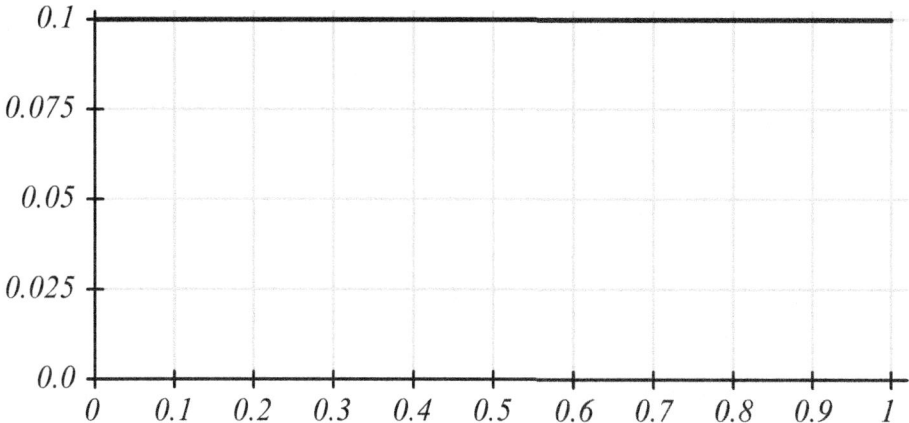

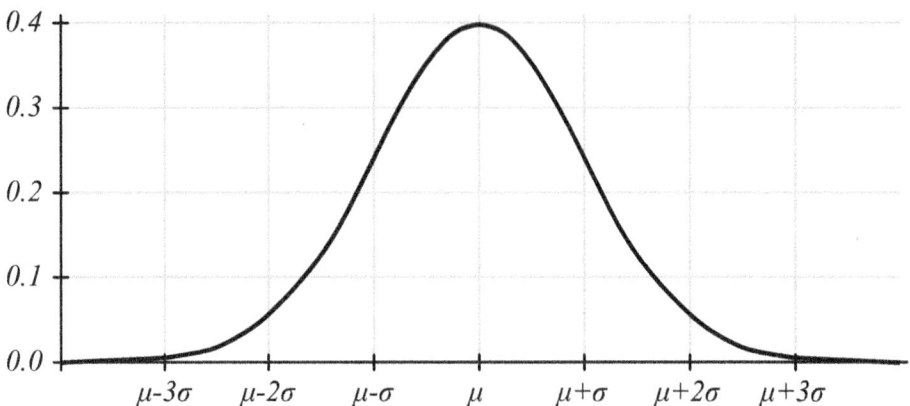

## 7.2 Probability Distributions

**Fig. 7.7** Histogram of random numbers created with the function (rand)

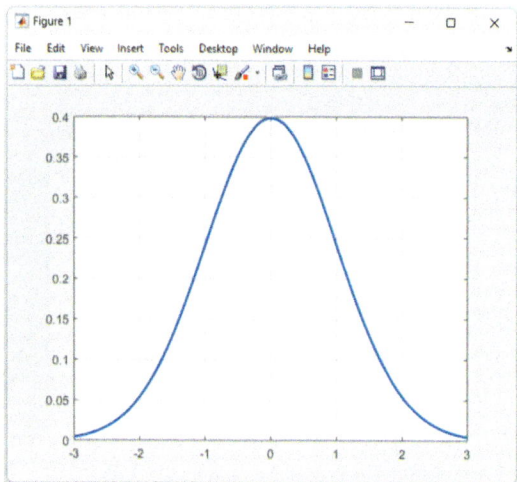

```
% Code 7.1 - Plot normal distribution by evaluating equation
clear ; clc ; close ;    % reset

% Parameters of the standard normal distribution
mu = 0 ;                 % Mean
sigma = 1 ;              % Standard deviation

x = linspace(-3, 3, 1000) ;
pdf = (1/(sigma*sqrt(2*pi))) * exp(-(x-mu).^2 / (2*sigma^2)) ; %(7.1)
plot(x, pdf, 'LineWidth', 2); % plot
grid on;                 % turn on the grid
```

If everything goes well when executing these lines, the graph of the normal distribution will be shown, as in Fig. 7.7.

The following code 7.2 is an exercise in which we generate a million random numbers with the function (rand) and subsequently plot them in a histogram.

```
% Code 7.2 - Exercises (rand)
clear ; clc ; close ;    % reset

data = rand(1000000, 1) ;
plot = histogram(data, 1000) ; % 1000 bins (intervals)
```

The first line of Code 7.2 generates a vector of 1 million elements with the (rand) function. This function, without arguments (no numbers in parentheses), generates a scalar randomly. If we add arguments, it generates a matrix of elements selected randomly. To generate a vector of one million elements, we can do so as a one million by one matrix.

**Fig. 7.8** Histogram of random numbers created with the (rand) function

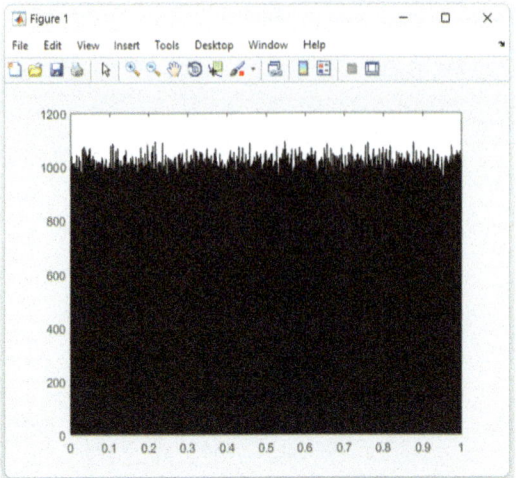

If everything goes well, executing Code 7.2 will display the graph of Fig. 7.8. We can observe that there are roughly a thousand results in each of the thousand intervals (bins), dividing the million generated numbers almost uniformly.

We can see that Fig. 7.8 is very similar to Fig. 7.5. However, Fig. 7.8 has some variations due to the randomness of the experiment.

Now let's look at the histogram of the (randn) function, a similar experiment to the previous one, but with a normal distribution, which can be carried out with Code 7.3.

```
% Code 7.3 - Exercises (randn)
clear ; clc ; close ; % reset

data = randn(1000000, 1) ;
plot = histogram(data, 1000) ; % 1000 bins (groups)
```

If everything goes well, the Matlab graph should show a result like that of Fig. 7.9. The numbers have also been grouped into 1000 intervals, with the most populated intervals near zero, with roughly 4 thousand elements each. The farther away from zero, the less populated the groups are.

We can now notice that Fig. 7.9 is very similar to the graphs of the normal distribution, like Figs. 7.4, 7.6, and 7.7.

In short, the (randn) function generates random numbers, and it is most likely that these numbers will be between − 1 and 1, but they could also have a greater magnitude, for example, between − 2 and 2. Theoretically, it can generate numbers of large magnitude, like − 100, 250, etc. But the probability of this happening is very low.

**Fig. 7.9** Histogram of random numbers created with the (randn) function

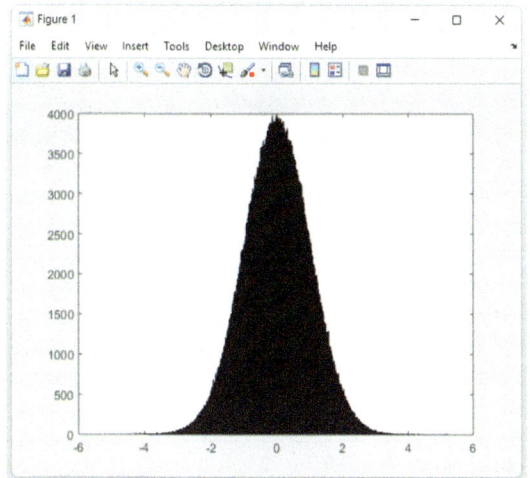

## 7.3 The Random Search Algorithm

The Random Search method was used in John's problem with volume discounts. The random search method, like many optimization algorithms, has variants, methods with modifications that seek to improve various aspects of the original algorithm. We will review a couple of variants of Random Search that may be useful [7–9]. Additionally, they will help understand how a method can evolve with changes in its search algorithm.

Now let's look at the random search method in its "Local Random Search" variant. The goal is to start from a solution to test, and instead of searching for the next one randomly across the entire variable range, it searches for a solution with a higher probability in the vicinity or closeness of the current best solution.

Suppose we want to find the highest place on a mountain, see Fig. 7.10. We know that the search range is between 0 and 7 km from the origin. Furthermore, suppose that this maximization is performed with the Random Search algorithm. This algorithm searches (in its simplest form) for a solution with a uniform distribution. And it keeps the best solution obtained up to the current iteration. This means the next solution to evaluate could be anywhere within the dependent variable with the same probability, assuming that the best solution so far is 1 (see Fig. 7.10), the next solution to try could be 2.1, 3.5, 4.8, ..., 6.3, with the same probability.

Because the probability distribution in the random search method is uniform, we use the (rand) function.

The local random search method [10] uses a normal distribution, with a mean at the current position, see Fig. 7.11. Like the previous algorithm, the newly generated random solution is evaluated, and the best solution is kept. If the new solution is better, it is saved; otherwise, it is discarded.

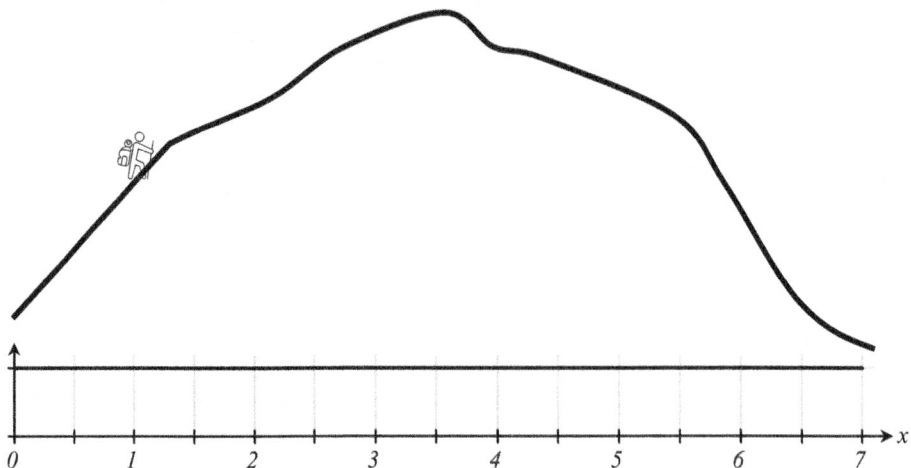

**Fig. 7.10** Maximization problem with the random search method (rand)

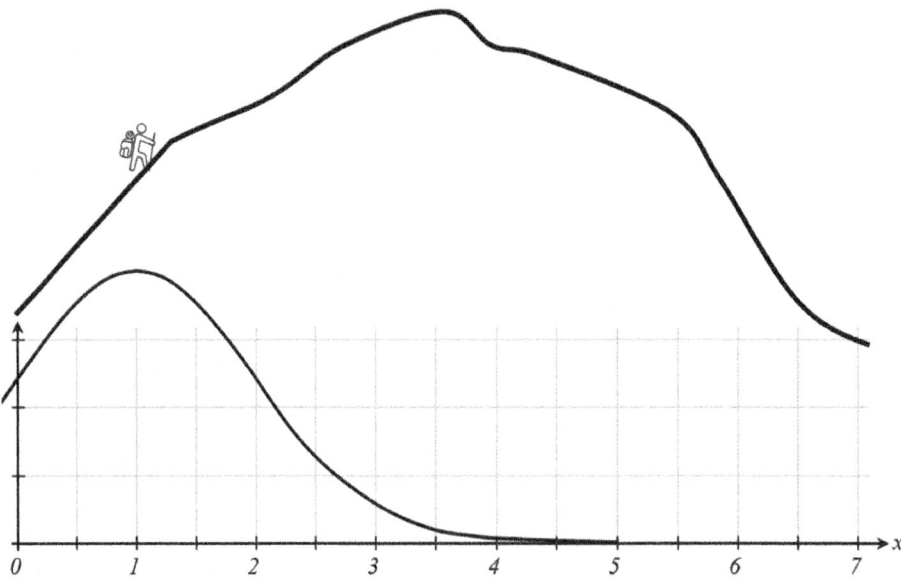

**Fig. 7.11** Maximization problem with the local random search method (rand)

But the distribution is normal, with a distribution centered on the best solution so far, see Fig. 7.11. This means the next solution is most likely to be in the vicinity of the current solution.

## 7.3 The Random Search Algorithm

**Table 7.1** Parameters of (7.2)

| Demand $d$ | $d = 5$ galon per day |
|---|---|
| Transportation cost $k$ | $k = 10$ USD per trip to the village |
| Storage cost $h$ | $h = 0.05$ USD per gallon per day |
| Cost per gallon $P$ | $P = 1$ USD per gallon if $0 < Q < 120$<br>$P = 0.75$ USD per gallon if $120 \leq Q < 160$<br>$P = 0.60$ USD per gallon if $160 \leq Q \leq 200$ |

Recalling the explanation from Fig. 7.4 and the parameters of the normal distribution, assuming a standard deviation of 1, and a mean of 1 (as in Fig. 7.11), the probability that the next solution is between 0 and 2 is 68%, and the probability that the answer is between $-1$ and 3 is 95%. Remember, the range of the independent variable ($x$) in Fig. 7.11 is from 0 to 7, so having a solution at $-1$ is impossible, and we would have to discard such a solution.

The use of the normal distribution helps us focus the search in an area, however, it has the drawback that the solution can end up outside the range of the independent variable, so in algorithms that use a normal distribution, we invariably have to evaluate if the independent variable ($x$) is within the allowed range before evaluating the objective function.

Let's solve John's problem with volume discounts using the local random search method. As a reminder, the problem involves minimizing the following objective function.

$$\text{Min } f(Q) = \frac{d}{Q}k + \frac{Q}{2}h + dP. \tag{7.2}$$

$$\text{s.t. } 0 < Q \leq 200, \tag{7.3}$$

Considering the parameters in Table 7.1.

Code 7.4 solves the problem under study, which is John's problem with volume discounts, using the local random search algorithm.

```
% Code 7.4 - John's problem, with volume discounts, solved
% using the local random search method.
clear ; clc ; close ;    % reset

% Problem parameters
d = 5 ;              % daily demand in gallons
k = 10 ;             % cost to go to the village for the gallons
h = 0.05 ;           % daily storage cost per gallon

Px(   1:  119) = 1 ;        % Cost with quantity discounts
Px( 120:  159) = 0.75 ;
Px( 160:  200) = 0.60 ;

x = [ 1 : 1 : 200 ] ;    % generate an x axis (q)
CTr = k.*d./x ;          % Average transportation cost
CAl = x.*h./2 ;          % Average storage cost
CPr = d*Px ;             % Average product cost (gallons)

f = CTr + CAl + CPr ;

% Optimizer parameters
max_iter = 1000 ;    % maximum number of iterations
q = 20 ;             % initial point
n = 0 ;              % improvement counter
RStep = 50 ;         % Step size

for i = 1:max_iter   % iterations

    %q_n = 1 + round(199*rand) ; % generate a random step
    q_n = round(q + randn*RStep);

    if(( q_n >= 1 ) & ( q_n <= 200 ))

        if ( f(q_n) < f(q)) % evaluate if an update is needed
            q = q_n ;    % update

            n = n + 1 ; % increment the improvement counter
            [ n i q ]   % display the solution and counters

        end
    end


end

Solution = q
```

Let's now review the most important parts of Code 7.4, which is similar to the first random search code introduced in Chap. 2, but we'll go over it in detail.

The first lines of the code reset the memory, clear (delete) existing variables, clean the command window, and close open windows.

## 7.3 The Random Search Algorithm

```
% Code 7.4 - John's problem, with volume discounts, solved
% using the local random search method.
clear ; clc ; close ;   % reset
```

Subsequently, the program parameters are declared, including the cost of gallons with quantity discounts.

```
% Problem parameters
d = 5 ;                 % daily demand in gallons
k = 10 ;                % cost to go to the village for the gallons
h = 0.05 ;              % daily storage cost per gallon

Px(    1:  119) = 1 ;       % Cost with quantity discounts
Px(  120:  159) = 0.75 ;
Px(  160:  200) = 0.60 ;
```

Next, the vector of the independent variable is generated, which is actually $Q$, but for convention and familiarity, we call it "$x$-axis". We calculate the average costs of transportation, storage, and the product, and sum them up, which generates the objective function (7.2).

```
x = [ 1 : 1 : 200 ] ;   % generate an x axis (q)
CTr = k.*d./x ;         % Average transportation cost
CAl = x.*h./2 ;         % Average storage cost
CPr = d*Px ;            % Average product cost (gallons)

f = CTr + CAl + CPr ;
```

Then, the optimizer parameters are declared. We can notice that, in the optimizer parameters.

```
% Optimizer parameters
max_iter = 1000 ;    % maximum number of iterations
q = 20 ;             % initial point
n = 0 ;              % improvement counter
RStep = 50 ;         % Step size
```

After the improvement counter (n), the `RStep` parameter appears. We will use this parameter to adjust the step size, in this case (RStep = 50). We mentioned that the (randn) function will most likely generate a value between − 1 and 1, and it could even be between − 2 and 2. This randomly generated number will be multiplied by RStep, so, the steps in the search for a new solution will be in the range of + − 50, + − 00. It's a random parameter, but knowing it has a normal distribution, we can estimate the scope of the search.

Finally, the for loop of the optimization is executed.

```
for i = 1:max_iter    % iterations

    %q_n = 1 + round(199*rand) ; % generate a random step
    q_n = round(q + randn*RStep);

    if(( q_n >= 1 ) & ( q_n <= 200 ))

        if ( f(q_n) < f(q)) % evaluate if an update is needed
            q = q_n ;    % update

            n = n + 1 ; % increment the improvement counter
            [ n i q ]    % display the solution and counters

        end
    end

end

Solution = q
```

Within the (for) loop that executes the search algorithm, we generate the new point with the line of code (q_n = round(q + randn*RStep) ; ), telling us that the new q_n will be equal to the best solution q plus an increment obtained by multiplying the RStep parameter by the randn function. Remember that since the result of randn can be positive or negative, the algorithm can take a step forward or backward, but always around the best solution found so far (q). This is the main difference from the simple Random Search.

## 7.4 The Adaptive Random Search Method

Following this, before evaluating the function, we check if it is within the allowed range (if(( q_n> = 1) & ( q_n< = 200))) because the step could be large and exit the range of q, which must be between 1 and 200. Or perhaps the current best solution is close to 1, and if the step is negative, it could easily give us a negative q_n, which is a solution that should be discarded without evaluation.

If the randomly generated solution (though with the local random search method) is within the expected range for a solution, then we proceed to evaluate it and save the new solution if it is better than the previous best solution. The round instruction takes the integer part of the new solution, as the number of gallons to be purchased must be an integer.

An advantage of the local random search method is its ability to search around a particular point; its capacity for random jumps reduces the chances of getting trapped in a local optimum. However, if the RStep parameter is very small, the jumps around the current point could also be very small, which could reduce its chances of avoiding local optima.

## 7.4 The Adaptive Random Search Method

Another interesting variant of random search, which we will review in this chapter, is the adaptive search method [11]. The guideline of this method is that as iterations progress, the possible step size to search, RStep, gradually decreases. This allows for an initial exploration or extensive search, and later, as iterations advance, a meticulous search focused on an area or exploitation around where the best solution likely lies. This method is especially useful for objective functions with little variation in the region surrounding the global solution. In other words, when the gradient near the global optimum is small in magnitude.

Code 7.5 solves the problem under study, John's problem with volume discounts, using the adaptive random search method.

```matlab
% Code 7.5 - John's problem, with volume discounts, solved
% with the adaptive random search method.
clear ; clc ; close ;    % reset

% Problem parameters
d = 5 ;             % daily demand in gallons
k = 10 ;            % cost to go to the village for the gallons
h = 0.05 ;          % daily storage cost per gallon
P = 1 ;             % unit cost of the gallons

Px(   1: 119) = 1 ;       % Cost with quantity discounts
Px( 120: 159) = 0.75 ;
Px( 160: 200) = 0.60 ;

x = [ 1 : 1 : 200 ] ;    % generate an x axis (q)
CTr = k.*d./x ;          % Average transportation cost
CA1 = x.*h./2 ;          % Average storage cost
CPr = d*Px ;             % Average product cost (gallons)

f = CTr + CA1 + CPr ;

% Optimizer parameters
max_iter = 500 ;    % maximum number of iterations
q = 20 ;            % initial point
n = 0 ;             % improvement counter

for i = 1:max_iter    % iterations

    % Generate a random step adjusting with iteration
    q_n = round(q + randn*(100 - i/5));

    if(( q_n >= 1 ) & ( q_n <= 200 ))

        if ( f(q_n) < f(q)) % evaluate if an update is needed
            q = q_n ;    % update

            n = n + 1 ; % increment the improvement counter
            [ n i q ]   % display the solution and counters

        end

    end

end

Solution = q
```

## 7.5 Homework Example: Maximizing the Area with Fixed Perimeter

The main differences between Code 7.5 and Code 7.4 are that in the new Code 7.5, the parameter (RStep) does not appear, and when searching for a new random solution, the function (rand) is multiplied by the factor (100 - i/5)). The variable i keeps track of the iterations, and we can note that initially, the mentioned factor (100 - i/5)) is practically 100, but as the iterations increase and approach the maximum number of iterations, which in this case is 500, the factor approaches zero. In other words, the search area is being reduced, ensuring that as iterations pass, the algorithm transitions from performing a search in a large area, what we normally call exploration in optimization, to a search in a smaller area, what we normally refer to as exploitation in optimization.

Upon executing the code, the speed and accuracy are very good, practically in all runs the algorithm yields the correct result, and if we increase the number of iterations, the reliability significantly increases.

As we progress through the chapters, we will address other more sophisticated algorithms, but in turn, based on a random search, which is the principle of random search. There are random algorithms designed for multimodal and multidimensional problems. We will revisit the problems addressed in Chaps. 1 and 2 to solve them with these variants of the random search method as illustrative examples.

## 7.5 Homework Example: Maximizing the Area with Fixed Perimeter

We will revisit some previously seen problems, starting with the problem of the farmer who plans to make a rectangular wooden enclosure (see Fig. 7.12) for his sheep. He had enough wood to fence a perimeter of 300 m (including the gate) and wanted to make the enclosure with the largest possible area.

The mathematical modeling carried out in Chap. 1 for this problem can be summarized as the perimeter not exceeding 300 m, which can be expressed as (7.4).

$$2X + 2Y = 300. \tag{7.4}$$

**Fig. 7.12** A rectangular area to be maximized

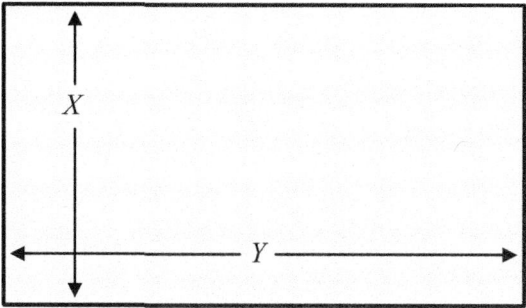

The area of the land which is desired to be maximized can be expressed as the product of $X$ by $Y$, see (7.5).

$$A = X \times Y. \tag{7.5}$$

To put the area as a function of $X$, we can solve for $Y$ from (7.4), resulting in (7.6).

$$Y = \frac{300 - 2X}{2}. \tag{7.6}$$

Subsequently, we can substitute (7.6) into (7.5). The area would be defined as (7.7).

$$A = X \left( \frac{P - 2X}{2} \right) = \frac{PX - 2X^2}{2} = X \frac{P}{2} - X^2. \tag{7.7}$$

So, the optimization problem is summarized in maximizing (7.7) subject to the restrictions described below:

$$\max_{X \in \mathbb{R}} f(X) = X \frac{P}{2} - X^2. \tag{7.8}$$

Subject to:

$$2X + 2Y = 300. \tag{7.9}$$

$$X > 0. \tag{7.10}$$

$$Y > 0. \tag{7.11}$$

Code 7.6 solves the described problem using the adaptive random search method:

## 7.6 Homework Example: Maximizing the Peaks Function

```
% Code 7.6 - Maximum area problem
% solved with the adaptive random search method.
clear ; clc ; close ;   % reset

P = 300 ;
f = @(x)  x*P/2 - x^2 ;       % Function

% Optimizer parameters
max_iter = 1000 ;    % maximum number of iterations
x = 10 ;             % initial point
n = 0 ;              % improvement counter

for i = 1:max_iter   % iterations

    x_n = x + randn*(100 - i/10);

    if(( x_n >= 0 ) & ( x_n <= 300 ))

        if ( f(x_n) > f(x))  % evaluate if update is needed
            x = x_n ;    % update

            n = n + 1 ;  % increment improvement counter
            [ n i x ]    % display the solution and counters

        end

    end

end

Solution = x
```

## 7.6 Homework Example: Maximizing the Peaks Function

Probablely, the benefits of stochastic algorithms become more apparent in a problem like the maximization of the Peaks function [12]. As mentioned, it is a multimodal and multi-dimensional function commonly used in numerical optimization courses and is described by Eq. (7.12).

$$f(x_i, x_2) = 3(1 - x_1)^2 \cdot e^{(-(x_1^2 - (x_2+1)^2))}$$
$$- 10\left(\frac{x_1}{5} - x_1^3 - x_2^5\right) \cdot e^{(-x_1^2 - x_2^2)} - \frac{1}{3}e^{(-(x_1+1)^2 - x_2^2)}. \qquad (7.12)$$

where both values of $x_1$ and $x_2$ are in the interval $(-3 \leq x_1 \leq 3)$, $(-3 \leq x_2 \leq 3)$. In Chap. 3, a code was presented to maximize the function using the gradient ascent algorithm. Repeated here for convenience.

```
% Code 3.8 - Optimize Peaks function with gradient
clear ; clc ; close ; % reset

f = @(x1,x2) 3*(1-x1).^2.*exp(-(x1.^2) - (x2+1).^2) ...
    - 10*(x1/5 - x1.^3 - x2.^5).*exp(-x1.^2-x2.^2) ...
    - 1/3*exp(-(x1+1).^2 - x2.^2) ;

% Create a grid of points in space
[x1, x2] = meshgrid(linspace(-3, 3, 1000), linspace(-3, 3, 1000)) ;
z = f(x1, x2) ; % Evaluate the function at each point on the grid

figure('Position', [100 100 1200 400]) ;
subplot(1, 2, 1) ;
surf(x1, x2, z, 'EdgeColor', 'none', 'FaceColor', 'interp') ;
hold on ;
colorbar ; % display color bar.
subplot(1, 2, 2) ;
contour(x1,x2,z,20) ;
hold on ;

% Parameters for the gradient method
alpha = 0.01 ; % Learning rate
Iter = 100 ; % Maximum number of iterations
p = [ -1; 2] ; % Starting point
h = 0.001 ;

for iter = 1:Iter % Iterative process

    p(1) = p(1) + alpha * (f(p(1)+h,p(2)) - f(p(1),p(2)))/h ;
    p(2) = p(2) + alpha * (f(p(1),p(2)+h) - f(p(1),p(2)))/h ;

    subplot(1, 2, 1) ;
    plot3(p(1), p(2), f(p(1),p(2)), 'ro', 'MarkerSize', 2 ) ;
    subplot(1, 2, 2) ;
    plot(p(1),p(2),'.','markersize',10,'markerfacecolor','g') ;
    pause(0.05) ;

end

Sol = p % Result coordinates of the highest point
Height = f(p(1),p(2)) % Height of the highest point
```

Something not mentioned is that the gradient algorithm can find the maximum of the peaks function when initialized in a region close to its highest peak, or at least, at a point whose gradient points in the direction of the highest peak, without encountering local optima that might trap the solution. For example, the previous code starts at the coordinates [-1, 2], but what happens if we initialize at points near other peaks, for example, at [-0.5, -1] or [2, 0]? The code is unable to find the maximum in this case; it gets trapped at one of the function's peaks.

Code 7.7 maximizes the peaks function with the adaptive random search method:

## 7.6 Homework Example: Maximizing the Peaks Function

```matlab
% Code 7.7 - Optimizing Peaks function with adaptive random search
clear ; clc ; close ;    % reset

f = @(x1,x2) 3*(1-x1).^2.*exp(-(x1.^2) - (x2+1).^2) ...
    - 10*(x1/5 - x1.^3 - x2.^5).*exp(-x1.^2-x2.^2) ...
    - 1/3*exp(-(x1+1).^2 - x2.^2) ;

% Create a grid of points in space
[x1, x2] = meshgrid(linspace(-3, 3, 1000), linspace(-3, 3, 1000)) ;
z = f(x1, x2) ; % Evaluate the function at each grid point

figure('Position', [100 100 1200 400]) ;
subplot(1, 2, 1) ;
surf(x1, x2, z, 'EdgeColor', 'none', 'FaceColor', 'interp') ;
hold on ;
colorbar ; % shows the color bar.
subplot(1, 2, 2) ;
hold on ;
contour(x1,x2,z,20) ;

% Parameters for the adaptive random search method
Niter = 1000 ; % Maximum number of iterations
p1 = -0.5 ;
p2 = -1 ;        % Starting point
n = 0 ;          % improvement counter

for iter = 1:Niter % Iterative process

    p_n1 = p1 + randn*3*(1 - iter/100) ;
    p_n2 = p2 + randn*3*(1 - iter/100) ;

    if( (p_n1>-3) && (p_n1<3) && (p_n2>-3) && (p_n2<3))

        if ( f(p_n1, p_n2) > f(p1,p2)) % check if update is needed

            p1 = p_n1 ;      % update
            p2 = p_n2 ;

            n = n + 1 ;      % increment improvement counter
            [ n iter f(p1,p2) ] % display status

            subplot(1, 2, 1) ;
            plot3(p1, p2, f(p1,p2), 'ro', 'MarkerSize', 2 ) ;
            subplot(1, 2, 2) ;
            plot(p1,p2,'.','markersize',10,'markerfacecolor','g') ;
            pause(0.05) ;

        end

    end

end

Solution = [p1 p2]       % Coordinates of the highest point result
Height = f(p1,p2)    % Height of the highest point
```

Upon executing the code, it is evident that even when the algorithm is initialized near a local optimum, it quickly transitions to the peak where the global optimum resides. This highlights the algorithm's exceptional speed, precision, and reliability.

Let's now dissect the code. Initially, the memory is reset, and the Peaks function is defined to allow evaluation with arguments $(x_1, x_2)$, which represent the coordinates of a solution to be tested.

```
% Code 7.7 - Optimizing Peaks function with adaptive random search
clear ; clc ; close ;   % reset

f = @(x1,x2) 3*(1-x1).^2.*exp(-(x1.^2) - (x2+1).^2) ...
    - 10*(x1/5 - x1.^3 - x2.^5).*exp(-x1.^2-x2.^2) ...
    - 1/3*exp(-(x1+1).^2 - x2.^2) ;
```

Subsequently, we create a figure that visually aids us in tracking how the algorithm progresses through the points of improvement it encounters. This process is explained in Chap. 2. A grid of coordinates is created, the objective function is evaluated across the entire grid, and subsequently, it can be visualized in 3D or 2D with contour assistance.

```
% Create a grid of points in space
[x1, x2] = meshgrid(linspace(-3, 3, 1000), linspace(-3, 3, 1000)) ;
z = f(x1, x2) ; % Evaluate the function at each grid point

figure('Position', [100 100 1200 400]) ;
subplot(1, 2, 1) ;
surf(x1, x2, z, 'EdgeColor', 'none', 'FaceColor', 'interp') ;
hold on ;
colorbar ; % shows the color bar.
subplot(1, 2, 2) ;
hold on ;
contour(x1,x2,z,20) ;
```

We then declare the parameters for the adaptive random search method.

```
% Parameters for the adaptive random search method
Niter = 1000 ; % Maximum number of iterations
p1 = -0.5 ;
p2 = -1 ;      % Starting point
n = 0 ;        % improvement counter
```

After setting the parameters, the for loop begins, which carries out the maximization process.

```
for iter = 1:Niter % Iterative process

    p_n1 = p1 + randn*3*(1 - iter/100) ;
    p_n2 = p2 + randn*3*(1 - iter/100) ;
```

## 7.6 Homework Example: Maximizing the Peaks Function

First, a new solution is generated randomly. It is notable that the solution is created randomly but starting from the previous point (p1, p1), which in the first iteration is the initial point. A randomly generated number is added to the initial point (+ `randn*3*(1 - iter/100)`), with the number multiplied by 3 so that the initial standard deviation is approximately 3. This ensures that the initial search covers the full function range. However, the factor multiplying the randomly generated number (`3*(1 - iter/100)`) gradually decreases as iterations progress. Thus, as iterations approach the maximum number (`Niter = 1000;`), the multiplier for the randomly generated value approaches zero.

Once the new solution is generated randomly but with determinable parameters, we verify if the new solution is within the objective function's range, from $-3$ to 3, for both x and y.

```
if( (p_n1>-3) && (p_n1<3) && (p_n2>-3) && (p_n2<3))
```

If the new solution is indeed within that range, we then assess if evaluating this new solution yields a higher objective function value than the previous solution:

```
if ( f(p_n1, p_n2) > f(p1,p2))  % check if update is needed
```

If so, it indicates that a new better solution has been found, prompting us to update the solution.

```
p1 = p_n1 ;      % update
p2 = p_n2 ;
```

In addition to updating the solution, we increment the improvement counter, a measure that informs us at the end how many better solutions were found throughout all iterations. We display the status on the screen containing the improvement counter, iteration number (helping us know at which of the thousand iterations the improvement was found), and the objective function evaluated at the new solution. This last number is expected to increase as new, better solutions are discovered.

```
n = n + 1 ;     % increment improvement counter
[ n iter f(p1,p2) ] % display status
```

Following this, we place two points on the plots, one for the 3D graphic and another for the 2D graphic.

```
              subplot(1, 2, 1);
              plot3(p1, p2, f(p1,p2), 'ro', 'MarkerSize', 2 );
              subplot(1, 2, 2);
              plot(p1,p2,'.','markersize',10,'markerfacecolor','g');
              pause(0.05);

         end

      end

   end
```

Finally, we display the solution, which is optional, as the improvement status shown on the screen contains the best solution found at the end.

```
      Sol = [p1 p2]          % Resultado coordenadas del punto más alto
      Altura = f(p1,p2)      % Altura del punto más alto
```

## 7.7 Homework Example: Minimizing the Bohachevsky Function

Finally, we will address the Bohachevsky function [4], a challenging optimization function. Even in Chap. 2, with the gradient descent algorithm, the global optimum (located at (0, 0)) was not achieved. The difficulty lies in its sinusoidal functions that create small waves (whose magnitude can indeed be adjusted through constant parameters). These waves generate local optima, making it more challenging to distinguish them in regions close to the global optimum since the gradient becomes smaller. The Bohachevsky function is described by Eq. (7.13).

$$f(x_i, x_2) = x_1^2 + 2 \cdot x_2^2 - 0.3 \cdot \cos(3 \cdot \pi \cdot x_1) - 0.4 \cdot \cos(4 \cdot \pi \cdot x_2) + 0.7. \quad (7.13)$$

The goal is minimization, i.e., to find the values of $x_1$ and $x_2$ that minimize (7.13). The values of the independent variables ($x_1$ and $x_2$) are within the range of ($-10 \leq \times 1 \leq 10$), ($-10 \leq \times 2 \leq 10$). In Chap. 3, a code was presented, implementing the gradient descent algorithm with the Bohachevsky function. It is repeated here for convenience.

## 7.7 Homework Example: Minimizing the Bohachevsky Function

```
% Code 3.10 - Optimizing the Bohachevsky Function with Gradient
clear; clc; close; % reset

f = @(x1,x2) (x1.^2) + 2.*(x2.^2) -0.3.*cos(3.*pi.*x1) ...
    -0.4.*cos(4.*pi.*x2) + 0.7;

% Create a grid of points in space
[x1, x2] = meshgrid(linspace(-10, 10, 1000), linspace(-10, 10, 1000));
z = f(x1, x2); % Evaluate the function at each grid point

figure('Position', [100 100 1200 400]);
subplot(1, 2, 1);
surf(x1, x2, z, 'EdgeColor', 'none', 'FaceColor', 'interp');
hold on;
colorbar; % shows the color bar
subplot(1, 2, 2);
contour(x1, x2, z, 20);
hold on;

% Parameters for the gradient method
alpha = 0.02; % Learning rate
Iter = 100; % Maximum number of iterations
p = [-8; -8]; % Starting point
h = 0.001;

for iter = 1:Iter % Iterative process

p(1) = p(1) - alpha * (f(p(1)+h,p(2)) - f(p(1),p(2)))/h;
p(2) = p(2) - alpha * (f(p(1),p(2)+h) - f(p(1),p(2)))/h;

subplot(1, 2, 1);
plot3(p(1), p(2), f(p(1),p(2)), 'ro', 'MarkerSize', 2);
subplot(1, 2, 2);
plot(p(1), p(2), '.', 'markersize', 10, 'markerfacecolor', 'g');
pause(0.05);

end

Solution = p; % Coordinates of the highest point
Height = f(p(1),p(2)); % Height of the highest point
```

The code, similar to the Bohachevsky function, generates a figure that includes both the 3D and 2D representations of the function. In the 2D function, it is evident that the solution does not lie at the center of the graph (0, 0).

Code 7.8 minimizes the Bohachevsky function using the adaptive random search algorithm.

```matlab
% Code 7.8 - Optimizes Bohachevsky function with adaptive random s
clear; clc; close all;    % reset

f = @(x1, x2) (x1.^2) + 2.*(x2.^2) - 0.3.*cos(3.*pi.*x1) ...
              - 0.4.*cos(4.*pi.*x2) + 0.7;

% Create a grid of points in space
[x1, x2] = meshgrid(linspace(-10, 10, 1000), linspace(-10, 10, 1000));
z = f(x1, x2);    % Evaluate the function at each grid point

figure('Position', [100 100 1200 400]);
subplot(1, 2, 1);
surf(x1, x2, z, 'EdgeColor', 'none', 'FaceColor', 'interp');
hold on;
colorbar;    % display the color bar
subplot(1, 2, 2);
contour(x1, x2, z, 20);
hold on;

% Parameters for the adaptive random search method
Niter = 10000;   % Maximum number of iterations
p1 = -8;
p2 = -8;         % Starting point
n = 0;           % improvement counter

for iter = 1:Niter    % Iterative process

    p_n1 = p1 + randn*10*(1 - iter/1000);
    p_n2 = p2 + randn*10*(1 - iter/1000);

    if ((p_n1 > -10) && (p_n1 < 10) && (p_n2 > -10) && (p_n2 < 10))

        if (f(p_n1, p_n2) < f(p1, p2))    % evaluate if it updates

            p1 = p_n1;    % update
            p2 = p_n2;

            n = n + 1;    % increment the improvement counter
            [n iter f(p1, p2)]    % display status

            subplot(1, 2, 1);
            plot3(p1, p2, f(p1, p2), 'ro', 'MarkerSize', 2);
            subplot(1, 2, 2);
            plot(p1, p2, '.', 'markersize', 10, 'markerfacecolor', 'g');
            pause(0.05);

        end
    end
end

Solution = [p1 p2]    % Coordinates of the highest point
Height = f(p1, p2)    % Height of the lowest point
```

Given the complexity of the function, a larger number of iterations (Niter = 10,000) was used. However, the computer performs the iterations in a very short time, achieving a solution very close to the global optimum in all executions. In the figure generated with the function in 3D and 2D, the algorithm's tendency to end in the central part of the image is evident.

We can also observe an advantage of metaheuristic algorithms: code 7.8 is very similar to code 7.7. Only the objective function and the optimizer parameters are updated. It is also important to note that since this program is used for minimization, the update of the new solution is made if the objective function in the new solution is lower than the previous one (not higher as in the case of maximization).

These last two examples have demonstrated the superiority of probabilistic algorithms over gradient-based algorithms when the problem presents multimodality and multidimensionality. The execution speed of the code and the reliability of finding the correct solution are very outstanding compared to gradient-based codes.

Now that we have witnessed the advantages of probabilistic algorithms, we are ready to explore some more modern and powerful probabilistic methods compared to the random search algorithm.

## References

1. Johnson, N. L., Kotz, S., & Balakrishnan, N.Continuous Univariate Distributions, Volume 1 (2nd Edition), Continuous Univariate Distributions, Volume 1 (2nd Edition), 1994.
2. Rice, J. A. Mathematical Statistics and Data Analysis (3rd Edition). Cengage learning, 2006.
3. Kaelo, P., Ali, M.M. Some Variants of the Controlled Random Search Algorithm for Global Optimization. J Optim Theory Appl. 2006, 130, 253-264.
4. Mairaj, A., Al Bataineh, A., Kaur, D., & Javaid, A. Identifying the optimal solutions of Bohachevsky test function using swarming algorithms. In Proceedings on the International Conference on Artificial Intelligence (ICAI). 2019, 109–115.
5. Devore, J. L. Probability and Statistics for Engineering and the Sciences. 2008.
6. Walpole, R. E., Myers, R. H., and Myers, S. L. Probabilidad y estadística para ingenieros. Pearson educación. 1999.
7. Ali, M. M., Törn, A., AND Viitanen, S. A numerical comparison of some modified controlled random search algorithms. Journal of Global Optimization. 1997,11, 377–385.
8. Zhigljavsky, A. A. Theory of global random search, Springer Science & Business Media. 2012, 65.
9. Andradóttir, S., and Prudius, A. A. Adaptive random search for continuous simulation optimization. Naval Research Logistics (NRL). 2010, 57(6), 583-604.
10. Lourenço, H. R., Martin, O. C., and Stützle, T. Iterated local search. In Handbook of metaheuristics (pp. 320–353). Boston, MA: Springer US.2003.
11. Hamzaçebi, C., and Kutay, F. A heuristic approach for finding the global minimum: Adaptive random search technique. Applied Mathematics and Computation. 2006, 173(2), 1323-1333.
12. Jones, D. R., Schonlau, M., and Welch, W. J. Efficient global optimization of expensive black-box functions. Journal of Global optimization. 1998, 13, 455-492.

# The Simulated Annealing Method

## 8.1 Introduction

Simulated Annealing is a metaheuristic optimization technique inspired by the physical annealing process used in metallurgy. This method has become a fundamental tool in the field of computational optimization due to its ability to find approximate solutions to complex problems, especially those involving a large number of variables where traditional methods may fail or be inefficient. It is one of the best-established methods in the field of single-particle methods.

The concept of Simulated Annealing was introduced in the 1980s by S. Kirkpatrick, C. D. Gelatt, and M. P. Vecchi [1], although its roots could be traced back to earlier works in statistics and physics. The technique is named after the annealing process in metallurgy, where a material is heated and then cooled slowly to decrease its internal energy and increase its strength and ductility. Similarly, Simulated Annealing seeks to optimize an objective function (comparable to the system's internal energy) through a controlled exploration of the solution space, occasionally allowing moves that do not improve the current solution in order to avoid getting trapped in local optima.

In the optimization context, Simulated Annealing stands out for its versatility and robustness, being applicable to a wide range of problems, from optimizing logistics routes to the arrangement of components in electronic circuits. Unlike gradient-based methods, its structure decreases the likelihood of the solution getting trapped in local optima or suboptimal solutions. Simulated Annealing explores the solution space more thoroughly, increasing the probability of finding global or near-global optimal solutions.

The goal of Simulated Annealing is to explore considering the "cooling" approach, trying to emulate the cooling of metals [2, 3]. Initial decisions allow significant changes, enabling a process of exploration or extensive search in a large area, but as the "system" cools down, changes become more subtle and directed, allowing a process of exploitation

## 8.2 Description of the Simulated Annealing Method

Similar to other single-particle methods, such as gradient-based or random search methods, the Simulated Annealing algorithm starts with an initial solution $x_{actual}$ that can be chosen randomly. A parameter called the initial temperature $T_0$ is defined; theoretically, the initial temperature is high and will cool down as iterations pass, and this cooling will gradually reduce the search area, allowing the transition from exploration to exploitation [4–6].

In each iteration, a new solution $x_{new}$ is generated from the current solution plus a random perturbation; the perturbation is generated with a normal distribution multiplied by the current temperature.

$$x_{nueva} = x_{actual} + randn \times T_{actual}. \tag{8.1}$$

If the new solution is within the allowed range for the independent variable, the new solution is evaluated. For example, if we expect a positive number and (8.1) gives us a negative number, we discard that iteration's solution and move to the next iteration.

If the new solution is a valid solution (it is within the range of the independent variable), we evaluate it in the objective function $f(x)$. And we calculate what in this method is known as the cost difference or energy difference ($\Delta E$), which is simply the difference between the objective function evaluated with the new solution minus the objective function evaluated with the previous solution.

$$\Delta E = f(x_{nueva}) - f(x_{actual}). \tag{8.2}$$

Considering minimization. If $\Delta E$ is less than zero, this means that the new solution is better than the current solution, then the new solution becomes the current solution.

However, there is another condition in which the new solution becomes the current solution (even if it's worse) if a random statement is satisfied:

$$e^{\left(\frac{-\Delta E}{T_{actual}}\right)} > rand. \tag{8.3}$$

On the left side of (8.3), we have the expression of $e$ (the Euler's number 2.7183), raised to a fraction containing the negative of the energy difference ($\Delta E$) divided by the temperature. On the right side, we have the rand function, which will provide a random value between 0 and 1, with a uniform distribution.

Suppose the initial temperature is 1. In the early iterations, we would practically have $e$ raised to the negative of the energy difference. For solutions worse than the current solution, the energy difference is negative, so it would be a positive number.

So, there are two scenarios for updating the solution: the first is when the new solution is better than the previous one, meaning the energy difference is positive (see 8.2). The second scenario occurs when (8.3) is met because a positive energy difference would result in the solution being updated. We can focus the analysis of (8.3) on cases where the energy difference is positive.

Initially, it's very likely that (8.3) will be fulfilled, as we discussed. Suppose the initial temperature is 1, and the new solution is worse than the previous one, making the energy difference 1. This would make the exponential function approximately 2.7183, while the random function can generate values between 0 and 1. This results in several exploration jumps at the start of the algorithm, covering a relatively large area of the search space. Later, as the temperature decreases, the exponent becomes smaller, reducing the likelihood of fulfilling (8.3). At this stage, exploration ends, giving way to exploitation, where the new solution must practically be better to update the current solution. During the initial exploration, it's possible to find the best solution, which may be replaced due to the exploratory policy and probabilities. However, the algorithm considers keeping the best solution found along the way, which could be the ultimate solution.

Lastly, we can mention the equation used for cooling or reducing the temperature. Each iteration updates the temperature using Eq. (8.4), where the current temperature is multiplied by a factor $\alpha$ (less than 1), causing the temperature to decrease slightly with each iteration.

$$T_{nueva} = \alpha \times T_{actual}. \tag{8.4}$$

The algorithm can be stopped in several ways, such as by setting a maximum number of iterations or by defining a minimum temperature. Once the minimum temperature is reached, the algorithm stops. The algorithm is widely used in various engineering applications [7, 8].

## 8.3 Example of the Simulated Annealing Method

Code 8.1 solves John's problem using the simulated annealing algorithm. Let's execute the code before going over its details.

```matlab
% Code 8.1 - Problem 1.0, simulated annealing
clear; clc; close;   % reset

f = @(x) 50./x + 0.05*x./2 + 5 ;      % Function

% Initial parameters
x0 = 10 ;                 % Starting point
T = 1 ;                   % Initial temperature
alpha = 0.9999 ;          % Cooling factor
max_iter = 4000 ;         % Maximum number of iterations per temperature

% Initialization
x_current = x0 ;
f_current = f(x_current) ;
x_best = x_current ;
f_best = f_current ;

% Main loop of the algorithm
for j = 1:max_iter
    % Perturb the current solution (ensuring x_new is positive)
    x_new = abs(x_current + randn * T);
    f_new = f(x_new);

    % Acceptance criterion
    if (f_new < f_current) || (exp((f_current - f_new)/T) > rand())
        x_current = x_new;
        f_current = f_new;

        % Update the best found
        if f_new < f_best
            x_best = x_new;
            f_best = f_new;
        end
    end

    T = T * alpha ;
end

% Display results
Sol_best = x_best
Cost_best = f_best
```

Just like the previous codes, Code 8.1 begins by resetting the memory and then declares the objective function and initial parameters.

```matlab
% Code 8.1 - Problem 1.0, simulated annealing
clear; clc; close;   % reset

f = @(x) 50./x + 0.05*x./2 + 5 ;      % Function

% Initial parameters
x0 = 10 ;                 % Starting point
T = 1 ;                   % Initial temperature
alpha = 0.9999 ;          % Cooling factor
max_iter = 4000 ;         % Maximum number of iterations per temperature
```

The initial point can be calculated with a probabilistic function to make it random, though in this case, we are starting at ($x_0 = 10$). The initial temperature is 1, alpha (the cooling factor) is 0.9999, and the number of iterations has been set to 4000.

Afterwards, the main loop of iterations starts, the first lines define the new solution and evaluate this solution against the objective function.

```
% Main loop of the algorithm
for j = 1:max_iter
    % Perturb the current solution (ensuring x_new is positive)
    x_new = abs(x_current + randn * T);
    f_new = f(x_new);
```

Now we evaluate the acceptance of the new solution, which depends on two possible options: whether the solution is better, or whether Eq. (8.3) is satisfied.

```
    % Acceptance criterion
    if (f_new < f_current) || (exp((f_current - f_new)/T) > rand())
        x_current = x_new;
        f_current = f_new;
```

Next, we check if the new solution is better than the best one found so far, and if so, it gets updated. After this, the temperature is updated with Eq. (8.4), and the iteration ends. At the end of all iterations, the best solution is displayed.

```
        % Update the best found
        if f_new < f_best
            x_best = x_new;
            f_best = f_new;
        end
    end
    
    T = T * alpha ;
end

% Display results
Sol_best = x_best
Cost_best = f_best
```

## References

1. Kirkpatrick, S., Gelatt Jr, C. D., and Vecchi, M. P. Optimization by simulated annealing. science. 1983, 220(4598), 671–680.
2. Dekkers, A., & Aarts, E. Global optimization and simulated annealing. Mathematical programming. 1991, 50, 367-393.
3. Brooks, S. P., and Morgan, B. J. Optimization using simulated annealing. Journal of the Royal Statistical Society Series D: The Statistician. 1995, 44(2), 241-257.

4. Nikolaev, A. G., and Jacobson, S. H. Simulated annealing. Handbook of metaheuristics. 2010, 1–39.
5. Jansen, T. Simulated annealing. In Theory Of Randomized Search Heuristics: Foundations and Recent Developments. 2011, 171–195.
6. Guilmeau, T., Chouzenoux, E., and Elvira, V. Simulated annealing: A review and a new scheme. In 2021 IEEE Statistical Signal Processing Workshop (SSP). 2021, 101–105, IEEE.
7. Gonzalez-Ayala, P., Alejo-Reyes, A., Cuevas, E., & Mendoza, A. (2023). A Modified Simulated Annealing (MSA) Algorithm to Solve the Supplier Selection and Order Quantity Allocation Problem with Non-Linear Freight Rates. Axioms, 12(5), 459.
8. Gonzalez-Ayala, P., Alejo-Reyes, A., Cuevas, E., & Mendoza, A. A Modified Simulated Annealing (MSA) Algorithm to Solve the Supplier Selection and Order Quantity Allocation Problem with Non-Linear Freight Rates. Axioms. 2023, 12(5), 459.

# The Particle Swarm Optimization Method

## 9.1 Introduction

So far, optimization techniques that are based on having an initial or current solution and iteratively searching for a better solution than the previous one have been studied. When a better solution is found, the previous one is discarded and replaced with the new better solution. That is, there's always one solution, the best one up to the moment, and although there might be other solutions, we don't keep them; we erase them from memory replacing them with the new best one as soon as it appears.

Research in optimization methods has techniques that work with multiple solutions (not just one), trying to use the information from multiple solutions to perform more sophisticated optimization. For example, instead of keeping only the best solution, the 5 best or 10 could be kept.

Techniques that have several solutions stored (instead of one) are called multiple particle techniques, where each particle is a solution. In this scenario, the techniques we have previously studied, such as gradient descent or random search, are called one-particle techniques.

In the context we will study, it is said that a particle moves, which means it is eliminated and replaced with another solution or particle. The method of gradient descent sounds intuitive, in which the new solution is determined by the gradient of the objective function from the previous solution. This can be described as a movement, like the movement of a particle, which remains in motion until it reaches its final position, which is considered the obtained solution. When executing the codes of the gradient descent chapter, those that included a graph, we could observe this apparent movement of a particle that goes up or down until finding the optimum, as the case may be.

This chapter will address Particle Swarm Optimization (PSO), a metaheuristic global optimization technique inspired by social behaviors observed in nature, particularly in flocks of birds or schools of fish. Initially developed by James Kennedy and Russell Eberhart in 1995 [1], PSO is based on the idea that coordinated movement and collective intelligence of a group can be used to find optimal solutions in a multidimensional search space.

PSO is not only an effective and widely used optimization technique but has also started a trend among optimization method researchers who have developed more techniques attempting to emulate animal behaviors in nature.

In PSO, there are multiple solutions; each individual solution within the search space is represented by a "particle." Each particle moves through the search space with a velocity that is individually adjusted based on the particle's experience and that of its neighboring particles. The movement of each particle is influenced both by the best position it has found (personal cognition) and by the best position found by the swarm (social cognition). This balance between exploration (searching for new areas) and exploitation (deepening in known areas) allows the swarm to converge toward optimal solutions.

PSO has become a popular and effective tool for solving complex and nonlinear optimization problems in various fields [2, 3]. Its simplicity in terms of concepts and easy implementation make it attractive for many practical applications. Moreover, PSO does not require the optimization problem to be differentiable, similar to the random search method and in contrast to gradient-based methods. This makes it particularly useful in scenarios where traditional methods fail or are difficult to apply.

The flexibility of PSO has allowed its application across a wide range of areas, including, but not limited to, electrical engineering (for the design of electrical networks and control systems), bioinformatics (in DNA sequence alignment), industrial process optimization, and even in the financial sector for portfolio optimization. This broad applicability not only highlights the robustness of PSO but also its relevance in solving complex real-world problems.

## 9.2 Description of the PSO Method

As mentioned, PSO is inspired by the social and collective behavior of organisms such as birds and fish. In nature, these swarms exhibit behavior that can be described as collective intelligence; it seems that the group is capable of making complex decisions, when in reality, it's about individuals with their own thinking, without a central element of intelligence or a defined leader. This analogy is transferred to PSO, where a "swarm of particles" or a set of solutions to an optimization problem moves through the search space. Each "particle" in the swarm represents a possible solution and follows simple rules based on its experience and that of its neighbors to find local or global optima [4, 5].

In PSO, each particle has two main attributes: position and velocity [1, 6, 7]. The number of particles can be chosen as a parameter of the algorithm; we can imagine that

## 9.2 Description of the PSO Method

there are 10 particles, there could be 100, but to generalize we will talk about a number $i$ of particles.

The current position of a particle will be called $\mathbf{x}_{i(n)}$ and represents a potential solution in the problem's search space, its current velocity will be called $\mathbf{v}_{i(n)}$, and it determines how fast and in what direction the particle moves to explore new positions (solutions).

Remember that bold letters, like **x** and **v**, indicate that it is a vector; we can imagine the case of two dimensions, each vector has its "x" and "y" component.

In addition to individual position and velocity, there is a personal best position $\mathbf{p}_{best,i}$, and a global best position $\mathbf{g}_{best}$. This is because in PSO, the rule is that particles have to move, even if they are at the global optimum, movement is part of the algorithm. However, if a particle is forced to move, the information of the best solution is not lost; it is stored in that variable that indicates the best solution that the particle has had. In the end, it is possible to consult that solution and recover it in case it is the best solution.

Thus, the best solution of each particle $\mathbf{p}_{best,i}$ is kept. If i were equal to 10, indicating that we have 10 particles, we would have a $p_{best,1}, p_{best,2}, \ldots p_{best,9}$, and $p_{best,10}$. But only one $g_{best}$, which obviously would be one of those 10 best individual solutions $p_{best,i}$.

Now that we have defined these terms, we can add that the velocity of particle $i$, for the next movement (remembering that it is an iterative or step-by-step process), which we will call $\mathbf{v}_{i(n+1)}$, will be found through the following formula.

$$\mathbf{v}_{i(n+1)} = \omega \mathbf{v}_{i(n)} + c_1 r_1 \left(\mathbf{p}_{best,i} - \mathbf{x}_{i(n)}\right) + c_2 r_2 \left(\mathbf{g}_{best} - \mathbf{x}_{i(n)}\right). \tag{9.1}$$

Equation (9.1) indicates that the velocity of the next movement of particle i depends on its current velocity, or velocity from the previous movement vi(n), which is multiplied by a parameter or constant $\omega$, called the inertia factor, because it can regulate the impact that the previous velocity has on the new velocity.

In the next element of (9.1), we have the distance between the particle (in its current position $\mathbf{x}_{i(n)}$) and its best solution ($\mathbf{p}_{best,i}$). In the field of optimization, this distance ($\mathbf{p}_{best,i} - \mathbf{x}_{i(n)}$) is called personal cognition, this distance or cognition is multiplied by two factors: $c_1$ is a constant, called the personal acceleration constant, because it multiplies the distance of the particle with its best solution; $r_1$ is a random number between 0 and 1 ($[0, 1]$).

The last element of (9.1) contains the distance between the particle (in its current position $\mathbf{x}_{i(n)}$) and the best solution (of the $i$ particles) ($\mathbf{g}_{best}$). This distance ($\mathbf{g}_{best} - \mathbf{x}_{i(n)}$) is called social cognition, and is multiplied by two factors: $c_2$ is a constant, called the social acceleration constant, because it multiplies the distance of the particle with the best solution (of the entire society, so to speak), and $r_2$ is a random number between 0 and 1 ($[0, 1]$).

The constants $\omega$, $c_1$, and $c_2$, are selected by the programmer to adjust the impact each of the 3 factors has on the new velocity; $r_1$ and $r_2$ are random values. The factor of randomness remains present, as in most metaheuristic algorithms and in processes of nature.

Once defined how the velocity of the particles is calculated, we can define how the new position of the particles is calculated as in (9.2).

$$\mathbf{x}_{i(n+1)} = \mathbf{x}_{i(n)} + \mathbf{v}_{i(n+1)}. \tag{9.2}$$

In each iteration, particles adjust their velocities based on their personal experiences and those of the collective, thus exploring the search space. The balance between exploitation (focusing on known areas) and exploration (searching for new areas) is crucial for the effectiveness of PSO. Adjusting parameters such as $\omega$, $c_1$, and $c_2$, can have a significant impact on the swarm's behavior and, therefore, on the algorithm's efficiency in finding optimal solutions.

Before looking at a code, let's see a simple example. Figure 9.1 shows a contour diagram; in fact, it's the peaks function, which we studied in Chap. 3. Imagine we have 100 particles, but we'll only focus on one, the black particle. This particle was where the gray point is, with coordinates $(-2, -1)$, passed through where the green point is with coordinates $(0, -0.5)$, and now after 43 iterations, it is at the black point $(-1, 1)$. Let's assume it's particle $i = 5$, and since we're at iteration $n = 43$.

The current position ($n = 43$) of the particle ($i = 5$) can be written as (9.3).

$$\mathbf{x}_{i(n)} = \mathbf{x}_{5(43)} = \begin{bmatrix} -1 \\ 1 \end{bmatrix}. \tag{9.3}$$

The position of the best solution that particle 5 has had can be written as (9.4).

$$\mathbf{p}_{best,5} = \begin{bmatrix} -0.5 \\ 0 \end{bmatrix}. \tag{9.4}$$

And the best global position $\mathbf{g}_{best}$ can be written as (9.5).

$$\mathbf{g}_{best} = \begin{bmatrix} 0 \\ 2 \end{bmatrix}. \tag{9.5}$$

Let's also assume that the velocity $\mathbf{v}_{5(43)}$, of our particle is known as:

$$\mathbf{v}_{5(43)} = \begin{bmatrix} -0.1 \\ 0.15 \end{bmatrix}. \tag{9.6}$$

In the first iteration, the velocities can be considered zero, later, as iterations pass, the velocities are calculated with Eq. (9.1), in this case, we will calculate the velocity as follows.

$$\mathbf{v}_{5(44)} = \omega \mathbf{v}_{5(43)} + c_1 r_1 \left( \mathbf{p}_{best,5} - \mathbf{x}_{5(43)} \right) + c_2 r_2 \left( \mathbf{g}_{best} - \mathbf{x}_{5(43)} \right). \tag{9.7}$$

## 9.2 Description of the PSO Method

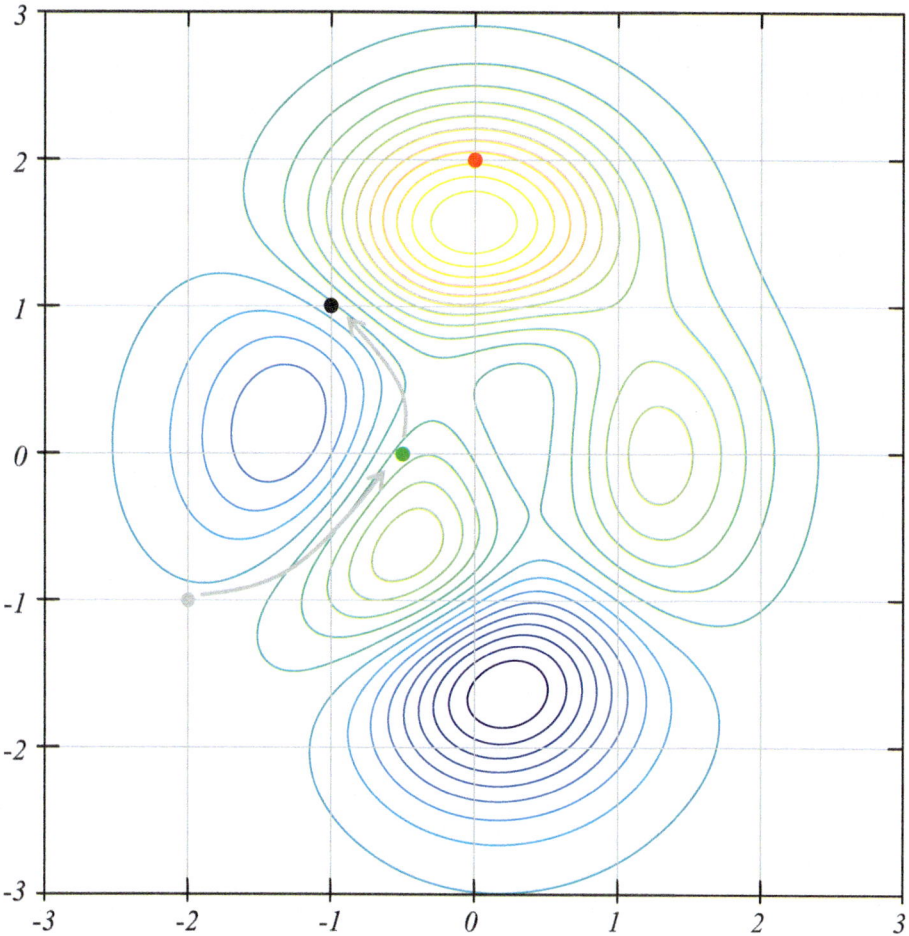

**Fig. 9.1** Example of a particle's behavior in PSO

Let's assume that $\omega = 0.5$, $c_1 = 1.5$, and $c_2 = 1.5$. And using the vectors we have defined in (9.3) to (9.6), Eq. (9.7) would be:

$$\mathbf{v}_{5(44)} = 0.5 \begin{bmatrix} -0.1 \\ 0.15 \end{bmatrix} + 1.5 r_1 \left( \begin{bmatrix} -0.5 \\ 0 \end{bmatrix} - \begin{bmatrix} -1 \\ 1 \end{bmatrix} \right) + 1.5 r_2 \left( \begin{bmatrix} 0 \\ 2 \end{bmatrix} - \begin{bmatrix} -1 \\ 1 \end{bmatrix} \right). \quad (9.8)$$

Note that the difference between the particle's best position and its current position, the personal cognition, would have 0.5 in its x component and -1 in its y component. It means that to go from the particle's current position to the best position it has had, it would have to advance 0.5 in the direction of the x-axis, and -1 in the direction of the y-axis, this would take us from the black point to the green point.

The social cognition would have 1 in its x component and 1 in its y component. This would take us from the black point to the red point. That's why vectors are sometimes named as distances.

Note that in Fig. 9.1, the red point, which is the best position obtained so far, is intentionally placed in a location different from the global optimum. It's assumed that the algorithm has not finished, but gradually, the particles could be getting closer to the global optima, which, according to what we studied in Chap. 3, would be approximately at (0, 1.5).

## 9.3 Useful Functions in Matlab

Before applying it, let's review some Matlab functions that will be highly useful in PSO and other metaheuristic algorithms.

If we have a matrix (A), we can select a column with the instruction A(:,i), where i is the column of interest. For instance, code 9.1 defines a matrix (A), then generates a column vector with column 1 of the matrix (A), and subsequently generates a row vector with row 2 of the matrix (A). Execute the code and observe the result in the command window.

```
% Code 9.1 - defines a matrix and calls a row or column
clear ; clc ; close ; % reset
A = [ 1 4 7 ; 2 5 8 ; 3 6 9 ]
B = A(:,1)
C = A(2,:)
D = C'
```

If all goes well, the command window should display matrix (A), its first column, and its second row. Note that the apostrophe can transpose a row vector and turn a row vector into a column vector, or vice versa (D = C').

We have already used the definition of functions in Matlab. For example, in code 9.2, a function (F1) is defined that adds a number to its square. Also defined are the $3 \times 3$ matrix (A) and the $2 \times 1$ vector (B).

## 9.3 Useful Functions in Matlab

```
% Code 9.2 - operations with matrices and functions
clear ; clc ; close ; % reset
F1 = @(x) x + (x.^2) ;
A = [ 1 4 7 ; 2 5 8 ; 3 6 9 ]
B = [ 1 ; 3 ]
R1 = F1(5)
R2 = F1(B)
R3 = F1(A(:,1))
```

In Matlab, there's no difference internally between vectors and matrices; vectors are a particular case of matrices.

Subsequently, in code 9.2, there's an example of evaluating a scalar with the function (F1(5)), evaluating a vector (F1(B)), and how to evaluate only a column of a matrix (F1(A(:,1))). Execute code 9.2 and observe the result to better understand functions with vectors and matrices.

There's a very useful Matlab function for finding the maximum values of a vector. Code 9.3 declares a vector (X), and then uses the function (max(X)) twice. The first time the function applies to the vector, it is assigned to a scalar (A = max(X)), and in that case, the function returns the maximum value of all values in the vector (A), which is 6.

```
% Code 9.3 - max function
clear ; clc ; close ; % reset
X = [ 1 ; 3 ; 5 ; 6 ; 4 ; 2 ]
A = max(X)
[A, B] = max(X)
```

The second time, the function applies to the vector in the same way, but this time, the function's response is assigned to an array of two values ([A, B] = max(X)). The result is that the first value stores the vector's maximum value (A = 6), and the second stores the element number in the vector (B = 4), i.e., its position within the vector, which is very useful to know the position of the maximum value.

Code 9.4 performs an exercise related to the peaks function. Let's observe it carefully.

```
% Code 9.4 - peaks max
clear ; clc ; close ; % reset
X = [ -2, -1 ; -0.5, 0 ; -1, 1 ; 0, 2]
pBestVal = peaks(X(:,1), X(:,2))
[gBestVal, gBestIdx] = max(pBestVal)
gBest = X(gBestIdx, :)
```

After resetting the memory. A vector (X) is declared, which contains 4 ordered pairs. These pairs are the coordinates of the points marked in Fig. 9.1, the gray point is [−2, −1], the green point is [−0.5, 0], the black point is [−1, 1], and the red point is [0, 2], knowing that this red point is the closest to the global maximum of this function, thus if we evaluate the coordinates in the peaks function, the red point should yield the largest number.

In Matlab notation, the semicolon means to move to the next row or line, thus all points can be declared in the same variable as an array of solutions.

Subsequently, code 9.2 evaluates the points with the function (peaks), and stores the value in the variable (pBestVal), which is indeed a vector of 4 elements, containing the height in peaks, of the points described in Fig. 9.1. Obviously, the largest value is the last one, as the red point is closest to the global optimum. Then, the code uses the function (max(pBestVal)) to obtain the maximum value and its index, finally, the pair of coordinates matching this value is stored in the variable (gBest). Execute the code and observe the results in the command window, which will help understand its operation.

In other words, we have a set of 4 coordinates (X), evaluate them in the peaks function, obtain the one that gives the highest value, and store both the highest value obtained from peaks (gBestVal), and the coordinates (gBest). Evidently, this set of lines will be used when solving the peaks function with PSO.

## 9.4 Maximizing the Peaks Function with the PSO Algorithm

Now, let's see how to apply the PSO algorithm to maximize the Peaks function. Code 9.5 performs this task; we will execute it and then discuss the code a bit.

## 9.4 Maximizing the Peaks Function with the PSO Algorithm

```matlab
% Code 9.5 - Peaks with PSO
clear ; clc ; close ;     % reset

% Create a grid of points in space
[x1, x2] = meshgrid(linspace(-3, 3, 1000), linspace(-3, 3, 1000)) ;
z = peaks(x1, x2) ;       % Evaluate the function at each grid point

figure('Position', [100 100 1200 400]) ; % Create figure frame
subplot(1, 2, 1) ;        % subplot 1
surf(x1, x2, z, 'EdgeColor', 'none', 'FaceColor', 'interp') ; % fig
hold on ;                 % prevent clearing
colorbar ;                % display color bar
subplot(1, 2, 2) ;        % subplot 2
contour(x1,x2,z,20) ;     % contour plot
hold on ;                 % prevent clearing

numPart = 100 ;           % Number of particles
X = rand(numPart, 2) * 6 - 3 ; % Initialize particles
V = zeros(numPart, 2) ;   % Initialize velocities

% Best personal and global positions
pBest = X ;
pBestVal = peaks(X(:,1), X(:,2)) ;
[gBestVal, gBestIdx] = max(pBestVal) ;
gBest = X(gBestIdx, :) ;

% PSO parameters
w = 0.5 ;                 % Inertia factor
c1 = 1.5 ;                % Cognitive coefficient
c2 = 1.5 ;                % Social coefficient
Niter = 200 ;             % Maximum number of iterations

for iter = 1:Niter % Iterative process

    % Update velocity
    V = w*V + c1*rand*(pBest - X) ...
        + c2*rand*(repmat(gBest, numPart, 1) - X) ;
    X = X + V ; % Update position

    for i = 1:numPart
        % Evaluate new position
        currentVal = peaks(X(i,1), X(i,2)) ;

        % Update (if applicable) the best personal position
        if currentVal > pBestVal(i)
            pBestVal(i) = currentVal ;
            pBest(i, :) = X(i, :) ;
        end
    end
```

```matlab
        % Update (if applicable) the best social (or global) position
        [maxPBestVal, idx] = max(pBestVal) ;
        if maxPBestVal > gBestVal
            gBestVal = maxPBestVal ;
            gBest = pBest(idx, :) ;

            subplot(1, 2, 1) ;
            plot3(gBest(1), gBest(2), peaks(gBest(1), gBest(2)), ...
                'ro', 'MarkerSize', 2) ;
            subplot(1, 2, 2) ;
            plot(gBest(1), gBest(2),'.','markersize',10, ...
                'markerfacecolor','g') ;
            pause(0.05) ;
        end

    end

    Solution = gBest % Result coordinates of the highest point
    Height = peaks(gBest(1), gBest(2)) % Height of the highest point
```

Let's now look at some details of the code. The program starts, as other codes we have studied, by resetting memory, creating the grid to plot the objective function, and plotting it. In this case, a figure with two sub-figures is made to see the peaks function in 3D and in 2D with the contour plot. In this case, the function is not defined to be evaluated; it uses the one defined in Matlab.

```matlab
% Code 9.5 - Peaks with PSO
clear ; clc ; close ;       % reset

% Create a grid of points in space
[x1, x2] = meshgrid(linspace(-3, 3, 1000), linspace(-3, 3, 1000)) ;
z = peaks(x1, x2) ;        % Evaluate the function at each grid point

figure('Position', [100 100 1200 400]) ; % Create figure frame
subplot(1, 2, 1) ;              % subplot 1
surf(x1, x2, z, 'EdgeColor', 'none', 'FaceColor', 'interp') ; % fig
hold on ;                       % prevent clearing
colorbar ;                      % display color bar
subplot(1, 2, 2) ;              % subplot 2
contour(x1,x2,z,20) ;           % contour plot
hold on ;                       % prevent clearing
```

Subsequently, the optimizer parameters are defined, and the vectors are initialized. In this case, 100 particles are being used (numPart = 100;). The position and velocity vectors are called X and V, respectively. They are, in fact, $100 \times 2$ matrices; however, each pair of scalars is considered a single velocity and position with its $x$ and $y$ components.

```matlab
numPart = 100 ;             % Number of particles
X = rand(numPart, 2) * 6 - 3 ; % Initialize particles
V = zeros(numPart, 2) ;     % Initialize velocities
```

## 9.4 Maximizing the Peaks Function with the PSO Algorithm

Note that the position vectors have been randomly initialized using the rand function, which, if used without arguments, returns a single value (a scalar) between 0 and 1. However, if arguments are placed in parentheses, for example, rand(M, n), it returns an (m x n) matrix with all values taken randomly between 0 and 1.

On the other hand, the velocities are initialized to zero with the zeros operator, which is used to generate matrices filled with zeros (V = zeros(numPart, 2);).

Once the first positions have been randomly generated, the following lines initialize the best positions both personally and socially. The best personal position is very simple since there is only one value for each particle, so the current solution is the best solution so far (pBest = X;).

Then the global best position is identified, within the 100 generated particles, we first have to evaluate the particles in the peaks function, then we can use the max function to save the height of the best solution, and the best solution (Yes, these are the same lines we studied previously).

```
% Best personal and global positions
pBest = X ;
pBestVal = peaks(X(:,1), X(:,2)) ;
[gBestVal, gBestIdx] = max(pBestVal) ;
gBest = X(gBestIdx, :) ;
```

Subsequently, the constant parameters of the PSO are declared, in this case, the inertia factor and cognitive coefficients, as well as the maximum number of iterations.

```
% PSO parameters
w = 0.5 ;              % Inertia factor
c1 = 1.5 ;             % Cognitive coefficient
c2 = 1.5 ;             % Social coefficient
Niter = 200 ;          % Maximum number of iterations
```

Now it is possible to start the iterative process that applies the PSO algorithm. Note that the update of velocity and position is very simple because MATLAB allows us to perform these functions in a matrix form.

```
for iter = 1:Niter % Iterative process

    % Update velocity
    V = w*V + c1*rand*(pBest - X) ...
            + c2*rand*(repmat(gBest, numPart, 1) - X) ;
    X = X + V ; % Update position
```

An interesting detail is that although the global best position (gBest) is a vector, it only contains one solution (it's a 1 × 2 vector). The other vectors are larger, containing the same number of solutions as the number of particles used, in this case, 100. A MATLAB function (repmat(gBest, numPart, 1)) helps us generate a vector that contains the solution (gBest) 100 times (numPart).

Within the iterative process, after updating the position of the particles, it's time to evaluate the new positions and update, if applicable, the best personal solution.

```matlab
for i = 1:numPart
    % Evaluate new position
    currentVal = peaks(X(i,1), X(i,2)) ;

    % Update (if applicable) the best personal position
    if currentVal > pBestVal(i)
        pBestVal(i) = currentVal ;
        pBest(i, :) = X(i, :) ;
    end
end
```

The possibility of updating the best global position is also evaluated, again if applicable, in other words, if a better global solution was found among the newly generated particles. In this case, it is also marked with a dot on the graphs, in other words, only the best global solution that is found is marked.

```matlab
    % Update (if applicable) the best social (or global) position
    [maxPBestVal, idx] = max(pBestVal) ;
    if maxPBestVal > gBestVal
        gBestVal = maxPBestVal ;
        gBest = pBest(idx, :) ;

        subplot(1, 2, 1) ;
        plot3(gBest(1), gBest(2), peaks(gBest(1), gBest(2)), ...
            'ro', 'MarkerSize', 2) ;
        subplot(1, 2, 2) ;
        plot(gBest(1), gBest(2),'.','markersize',10, ...
            'markerfacecolor','g') ;
        pause(0.05) ;
    end
end
```

Finally, the best solution is displayed, in its coordinates and in its height or value evaluated with the peaks function.

```matlab
Solution = gBest % Result coordinates of the highest point
Height = peaks(gBest(1), gBest(2)) % Height of the highest point
```

## 9.5 Minimizing the Bohachevsky Function

Now, let's look at Code 9.6, which applies the PSO algorithm, but this time not for maximization, but for minimization, and the objective function is the Bohachevsky function.

## 9.5 Minimizing the Bohachevsky Function

```matlab
% Code 9.6 - Bohachevsky with PSO
clear; clc; close;          % reset

Boha = @(x1,x2) (x1.^2) + 2.*(x2.^2) -0.3.*cos(3.*pi.*x1) ...
        -0.4.*cos(4.*pi.*x2) + 0.7 ;

% Create a grid of points in space
[x1, x2] = meshgrid(linspace(-10, 10, 1000), linspace(-10,10,1000)) ;
z = Boha(x1, x2) ;          % Evaluate the function at each grid point

figure('Position', [100 100 1200 400]) ;  % Create figure frame
subplot(1, 2, 1) ;          % Sub-figure 1
surf(x1, x2, z, 'EdgeColor', 'none', 'FaceColor', 'interp') ;  % 3D Fig
hold on ;                   % Do not clear
colorbar ;                  % Show color bar
subplot(1, 2, 2) ;          % Sub-figure 2
contour(x1,x2,z,20) ;       % Contour plot
hold on ;                   % Do not clear

numPart = 100 ;             % Number of particles
X = rand(numPart, 2) * 20 - 10 ; % Initialize particles
V = zeros(numPart, 2) ;     % Initialize velocities

% Personal and global best positions
pBest = X ;
pBestVal = Boha(X(:,1), X(:,2)) ;
[gBestVal, gBestIdx] = min(pBestVal) ;
gBest = X(gBestIdx, :) ;

% PSO Parameters
w = 0.5 ;                   % Inertia factor
c1 = 1.5 ;                  % Cognitive coefficient
c2 = 1.5 ;                  % Social coefficient
Niter = 100 ;               % Maximum number of iterations

for iter = 1:Niter          % Iterative process

    % Update velocity
    V = w*V + c1*rand*(pBest - X) ...
          + c2*rand*(repmat(gBest, numPart, 1) - X) ;
    X = X + V ;             % Update position

    for i = 1:numPart
        % Evaluate new position
        currentVal = Boha(X(i,1), X(i,2)) ;
        % Update personal best position (if applicable)
        if currentVal < pBestVal(i)
            pBestVal(i) = currentVal;
            pBest(i, :) = X(i, :);
        end
    end
```

```
            % Update global best position (if applicable)
            [maxPBestVal, idx] = min(pBestVal) ;
            if maxPBestVal < gBestVal
                gBestVal = maxPBestVal ;
                gBest = pBest(idx, :) ;

                subplot(1, 2, 1) ;
                plot3(gBest(1), gBest(2), Boha(gBest(1), gBest(2)), ...
                    'ro', 'MarkerSize', 2) ;
                subplot(1, 2, 2) ;
                plot(gBest(1), gBest(2),'.','markersize',10, ...
                    'markerfacecolor','g') ;
                pause(0.05) ;
            end

        end

        Sol = gBest  % Result coordinates of the highest point
        Altitude = Boha(gBest(1), gBest(2))  % Altitude of the lowest point
```

The main differences of this code compared to code 9.5 include the definition of the Bohachevsky function, as it's not predefined in Matlab. Additionally, given the objective here is minimization rather than maximization, the code employs the analog function min() instead of max(). Also, for updating the best solutions, the comparison checks if the new solutions are less than (<) the previous best, rather than greater than (>).

It's noteworthy that the PSO algorithm initializes particles randomly and, despite this, has a low likelihood of getting trapped in a local optimum, thanks to the number of particles and their movement dynamics.

## References

1. Kennedy J. and Eberhart, r. Particle Swarm Optimization. Proceedings of the IEEE International Conference on Neural Networks, 4, 1942–1948
2. Wang, D., Tan, D., & Liu, L. Particle swarm optimization algorithm: an overview. Soft computing. 2018, 22, 387-408.
3. Zhang, Y., Wang, S., and Ji, G. A comprehensive survey on particle swarm optimization algorithm and its applications. Mathematical problems in engineering. 2015.
4. Koohi, I., and Groza, V. Z. Optimizing particle swarm optimization algorithm. In 2014 IEEE 27th Canadian conference on electrical and computer engineering (CCECE). 2014, 1–5.
5. Sedighizadeh, D., and Masehian, E. Particle swarm optimization methods, taxonomy and applications. International journal of computer theory and engineering. 2009 1(5), 486-502.
6. Mazhoud, I., Hadj-Hamou, K., Bigeon, J., and Joyeux, P. Particle swarm optimization for solving engineering problems: a new constraint-handling mechanism. Engineering Applications of Artificial Intelligence. 2013, 26(4), 1263-1273.
7. Shami, T. M., El-Saleh, A. A., Alswaitti, M., Al-Tashi, Q., Summakieh, M. A., and Mirjalili, S. Particle swarm optimization: A comprehensive survey. IEEE Access. 2022, 10, 10031-10061.

# Evolutionary Strategies (ES)

## 10.1 Introduction

This chapter explains the function of the Evolution Strategies (ES) algorithm. It delves into the evolutionary process utilized by the ES method to tackle optimization problems. Additionally, the chapter outlines the various operators used in ES, their different variants, and their implementation in MATLAB.

In the 1960s, three students from the Technical University of Berlin, Ingo Rechenberg, Hans-Paul Schwefel, and Peter Bienert, faced a challenging problem they couldn't solve analytically. They decided to experiment with random solutions and measure their effectiveness. By selecting the best-performing random solutions, modifying them, and re-evaluating these modifications (referred to as mutations), they gradually improved their results. This iterative process continued until they achieved satisfactory outcomes.

Inspired by biological processes, where species evolve and adapt through mutations, the students theorized that solutions could similarly be enhanced by iterative mutations. This concept formed the foundation of the Evolution Strategies algorithm.

Notably, the initial development of this method was empirical due to limited computational resources. The students were attempting to optimize the shape of an aircraft wing to minimize drag in an airflow [1]. Different wing shapes were tested and evaluated experimentally.

The first version of the ES algorithm, known as $(1 + 1)$ ES [1, 2], involved a single parent solution that mutated to produce an offspring. If the offspring outperformed the parent, it replaced the parent; otherwise, it was discarded. This process repeated until a predefined stopping criterion was met.

Initially designed for discrete problems, this method often got stuck in local optima due to discrete mutations. To address this, Beyer and Schwefel modified the $(1 + 1)$ ES algorithm to function in continuous spaces [3].

Subsequent versions of the (1 + 1) ES method introduced additional mechanisms, such as recombination, alongside mutation. Some approaches also employed a population of solutions instead of a single parent–offspring pair, incorporating various forms of mutation, recombination, and selection.

This chapter will comprehensively detail the original ES method and its notable variants, including the adaptive (1 + 1) ES, ($\mu$ + 1) ES, ($\mu$ + $\lambda$) ES, ($\mu$, $\lambda$) ES, ($\mu$, $\alpha$, $\lambda$, $\beta$) ES, adaptive ($\mu$ + $\lambda$) ES, and adaptive ($\mu$, $\lambda$) ES.

## 10.2 The (1 + 1) ES

As previously mentioned, the (1 + 1) ES algorithm begins with a parent particle $x_p$ that is randomly initialized. This particle undergoes mutation to create a new particle $x_o$, which represents the offspring. If $x_o$ is superior to $x_p$, it replaces $x_p$; otherwise, it is discarded. Below is a more detailed explanation of this process.

### 10.2.1 Initialization

In the initialization phase of the (1 + 1) ES algorithm, a particle $x_p^0 = \{x_1^0, ..., x_d^0\}$ is randomly generated within the defined lower $lb$ and upper $ub$ bounds of the d-dimensional search space. The particle initialization is described by Eq. 10.1.

$$x_j^0 = lb_j + r_j(ub_j - lb_j)$$
$$j = 1, 2, ..., d. \quad (10.1)$$

Here, $x_j^0$ is the initial value of the particle in dimension $j$, $lb_j$ and $ub_j$ are the lower and upper bounds for each dimension, respectively, and $r_j$ is a random value between 0 and 1, typically generated using a uniform distribution, though other distributions can also be used.

In addition to initializing $x_p^0$, the algorithm defines the variance $\sigma^2$, a parameter crucial for mutating the parent. The value of $\sigma^2$ needs to balance exploration and exploitation: a high variance leads to excessive exploration and poor precision, while a low variance can cause the algorithm to get stuck in local optima and slow convergence.

### 10.2.2 Mutation

In the (1 + 1) ES method, the particle explores the solution space through mutations. Each element of $x_p$ is altered by adding a random value $\rho_j$ drawn from a normal distribution with a mean of zero and variance $\sigma^2$. The mutation is described by Eq. 10.2.

## 10.2 The (1 + 1) ES

$$x_o = x_p + \rho. \tag{10.2}$$

where $\rho = \{\rho_1, ..., \rho_d\}$ is a vector of normally distributed random values. Each element $\rho_j$ is defined in Eq. 10.3.

$$\rho_j = N(0, \sigma^2)$$
$$j = 1, 2, ..., d. \tag{10.3}$$

The mutation process is isotropic because the same variance $\sigma^2$ is used for mutating each element of $x_p$, ignoring covariance and resulting in a standard multivariate normal distribution with zero mean and covariance matrix $\Sigma$ defined in Eq. 10.4.

$$\Sigma = \begin{bmatrix} \sigma_1^2 & 0 & \cdots & 0 \\ 0 & \sigma_2^2 & \cdots & 0 \\ \vdots & \vdots & \ddots & \vdots \\ 0 & 0 & \cdots & \sigma_d^2 \end{bmatrix}. \tag{10.4}$$

$\Sigma$ is a symmetric $d \times d$ matrix, with the main diagonal comprising the variances $\sigma^2$ for each decision variable. Figure 10.1a illustrates a standard bivariate normal distribution, while Fig. 10.1b shows its top view.

The normal distribution in Fig. 3.1a appears circular because the variance is uniform across all decision variables. Under these mutation conditions, the offspring $x_o$ can be located anywhere within this circular region but is more likely to be near the center, close to $x_p$. Figure 10.2 depicts potential locations of the mutated particle around $x_p$.

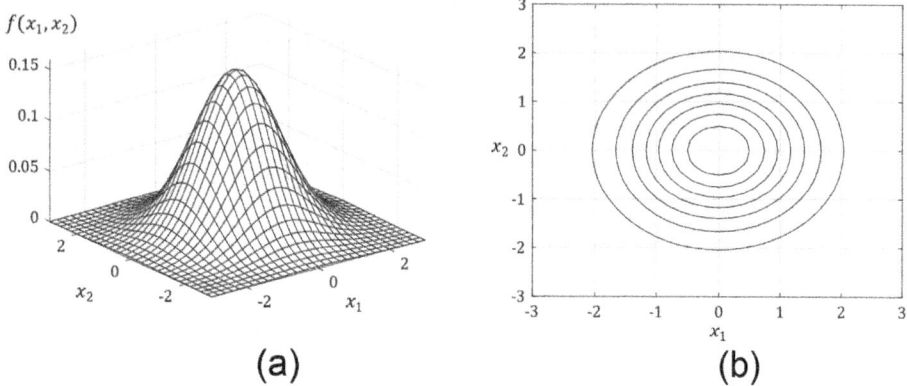

**Fig. 10.1** **a** A 3D view of the standard bivariate normal distribution and **b** its top view

**Fig. 10.2** Distribution after the mutation process

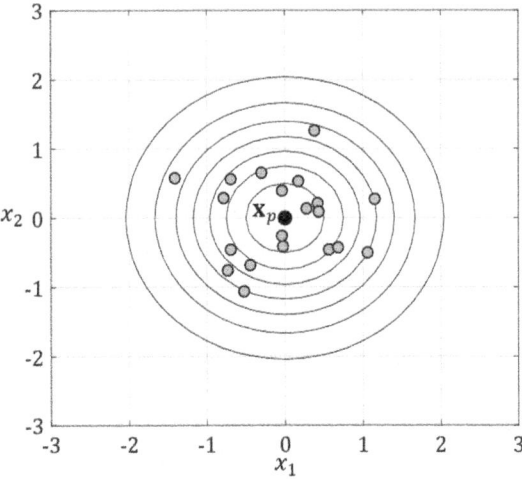

### 10.2.3 Selection

After mutation, the (1 + 1) ES algorithm uses an elitist selection method to choose the best particle for the next generation. This involves comparing $x_p$ and $x_o$ based on their fitness values. If $x_o$ has a better fitness value, it becomes the new $x_p$ for the next generation; otherwise, the original $x_p$ remains. The selection process is formalized as Eq. 10.5.

$$x_p^{(k+1)} = \begin{Bmatrix} x_p^k, & f(x_o^k) < f\left(x_p^k\right) \\ x_o^k, & f(x_o^k) \geq f\left(x_p^k\right) \end{Bmatrix}. \tag{10.5}$$

## 10.3 Simulation of the (1 + 1) ES

So far, the operators of the (1 + 1) ES algorithm have been outlined, and a general description of the method has been provided. In this section, we will delve deeper into the implementation and simulation details of the algorithm.

## 10.3 Simulation of the (1 + 1) ES

### 10.3.1 Methodology of the (1 + 1) ES Algorithm

The (1 + 1) ES algorithm begins by generating an initial particle $x_p^0$ randomly within the search space boundaries, as described by Eq. 10.1. Alongside this, the variance $\sigma^2$ is also initialized. The initial particle $x_p^0$ is then evaluated using the objective function to assess its quality as a solution.

Following the initialization, the iterative process starts. During each iteration, the mutation vector $\rho$ is created according to Eq. 10.3, which is then used to modify the particle $x_p^k$ based on Eq. 10.2, resulting in the offspring particle $x_o^k$. The offspring is then evaluated using the objective function. Finally, the best candidate is selected according to the criteria in Eq. 10.5. This cycle continues until a predefined stopping criterion is met. The flowchart of the (1 + 1) ES algorithm is depicted in Fig. 10.3.

The overall procedure of the method is outlined in Algorithm 10.1.

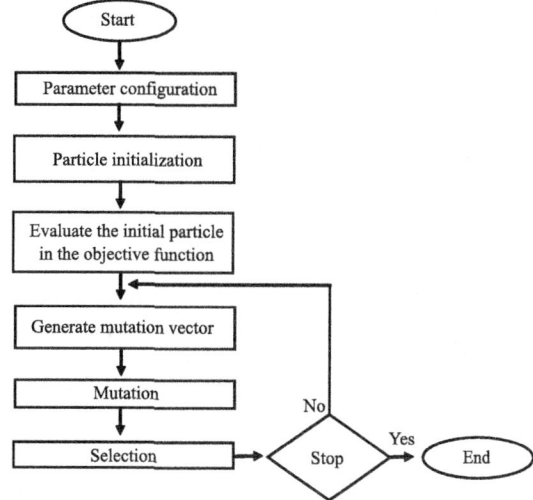

**Fig. 10.3** Flowchart of the (1 + 1) ES algorithm

---
**Algorithm 10.1**
The general procedure of the (1+1) ES algorithm.
---

Step 1: Parameter configuration
$$d, lb, ub, k \leftarrow 0, k_{max}, \sigma^2$$

Step 2: Particle initialization
$$x_p^k \leftarrow r \cdot (ub - lb) + lb$$

Step 3: Evaluation of the initial particle in the objective function
$$fx_p^k \leftarrow f(x_p^k)$$

Step 4: Generate mutation vector
$$\rho \leftarrow N(0, \sigma^2)$$

Step 5: Mutation
$$x_o^k \leftarrow x_p^k + \rho$$

Step 6: Evaluation of the offspring in the objective function
$$fx_o^k \leftarrow f(x_o^k)$$

Step 7: Selection
$$\text{if } fx_o^k < fx_p^k$$
$$x_p^k \leftarrow x_o^k$$
$$fx_p^k \leftarrow fx_o^k$$

Step 8: Verify the stop condition
$$\text{if } k = k_{max}$$
$$\text{Best solution} \leftarrow x_p^k$$
$$\text{End the search process}$$
$$\text{else}$$
$$\text{Go to Step 4}$$
$$\text{end}$$

---

By following these steps, the (1 + 1) ES algorithm systematically searches for the optimal solution within the given search space. The flowchart and Algorithm 10.1 provide a clear and structured overview of this process.

## 10.4 Implementation of the (1 + 1) ES Algorithm in MATLAB

The (1 + 1) ES algorithm has been implemented in MATLAB to illustrate its search procedure. This implementation is demonstrated in Code 10.1. In this example, the goal is to minimize the Rastringin objective function, which is defined by Eq. 10.6.

$$f(x) = 10d + \sum_{i=1}^{d} [x_i^2 - 10\cos(2\pi x_i)]. \tag{10.6}$$

with $x_i \in [-5.12, 5.12]$; $i \in \{1, ..., d\}$

Here, $d$ is the number of dimensions. The Rastringin function has a global optimum at $f(x^*) = 0$ where $x^* = \{0, ..., 0\}$. For $d = 2$, the Rastringin function reaches a global minimum of zero at $x_1 = 0$ and $x_2 = 0$. The shape of this function, which features a surface with multiple local optima, is depicted in Fig. 10.4.

Now, let's look at Code 10.1, which applies the (1 + 1) ES algorithm, where the objective function is the Rastringin function.

```
% Code 10.1
% Evolution Strategies algorithm (1+1) ES

% Clear memory and close windows
clear all
close all
% Parameter configuration
```

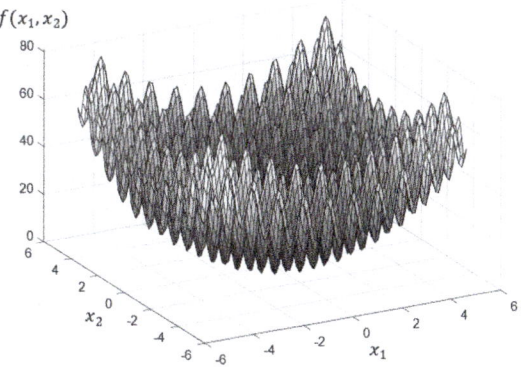

**Fig. 10.4** The Rastringin function in two dimensions

```matlab
    d = 2;                  % Number of dimensions
    lb = [-5.12 -5.12];     % Lower bound
    ub = [5.12 5.12];       % Upper bound
    k = 0;                  % Iteration counter
    kmax = 500;             % Maximum number of iterations
    sigma = 2;              % Standard deviation
    sigma2 = sigma^2;       % Variance
% Optimization problem (minimization), definition of the objective
function
    f = @(x) 10*d + sum(x.^2 - 10*cos(2*pi.*x));
% Particle initialization
    Xp = rand(1,d).*(ub-lb)+lb;
% Evaluation of the initial particle in the objective function
    fXp = f(Xp);
% Definition of the search space
    xAxis=linspace(min(lb),max(ub),85);
    yAxis=xAxis;
    zAxis=[];
    for i = 1:length(xAxis)
        for j = 1:length(yAxis)
            zAxis(i,j) = f([xAxis(i) yAxis(j)]);
        end
    end
    [yAxis,xAxis] = meshgrid(xAxis,yAxis);
% Iterative process
while k<kmax
    % Generate mutation vector
        rho = normrnd(0,sigma,1,d);
    % Mutation
        Xo = Xp + rho;
    % Verify search space limits
        for j=1:d
            if Xo(j) < lb(j)
                Xo(j) = rand*(ub(j)-lb(j))+lb(j);
            elseif Xo(j) > ub(j)
                Xo(j) = rand*(ub(j)-lb(j))+lb(j);
            end
        end
    % Evaluation of the offspring in the objective function
        fXo = f(Xo);
    % Draw the search space
        figure(1);
        surf(xAxis,yAxis,zAxis)
        hold on
    % Draw particles Xp and Xo

plot3(Xp(1),Xp(2),fXp,'o','MarkerFaceColor','m','MarkerSize',10)

plot3(Xo(1),Xo(2),fXo,'o','MarkerFaceColor','y','MarkerSize',10)
        pause(0.01)
        hold off
    % Draw the contour space
        figure(2)
        contour(xAxis,yAxis,zAxis,20)
        hold on
    % Draw Xp and Xo in the contour space
        plot(Xp(1),Xp(2),'o','MarkerFaceColor','m');
        plot(Xo(1),Xo(2),'o','MarkerFaceColor','y');
```

## 10.5 Variants of Evolutionary Strategies

```
                pause(0.01)
                hold off
            % Selection
                if fXo < fXp
                    Xp = Xo;
                    fXp = fXo;
                end
            k=k+1;
    end
    % Show results
            display(['Best solution: x1= ',num2str(Xp(1)),', x2= ',num2str(Xp(2))])
            display([' f(X)= ',num2str(fXp)])
```

This MATLAB code follows the $(1+1)$ ES algorithm to minimize the Rastringin function. The process begins with initializing a parent particle within the specified bounds. The particle undergoes mutation, generating an offspring that is then evaluated and compared to the parent. The better particle is selected for the next iteration, and this process continues until the maximum number of iterations is reached. By following this procedure, the algorithm searches for the optimal solution, navigating through the multiple local optima of the Rastringin function surface.

In the implementation described in Code 10.1, the Rastringin function is considered in 2 dimensions, with 500 iterations and a variance $\sigma^2 = 4$. Additionally, the code has been enhanced to visualize the search process. Upon execution of Code 10.1, the solution found by the $(1+1)$ ES algorithm approximates the expected solution. However, due to the complexity of the Rastringin function, the algorithm may not always reach or get close to the global minimum. This limitation has led to the development of other variants aimed at improving the algorithm's performance, achieving better solutions with greater accuracy in less time.

## 10.5 Variants of Evolutionary Strategies

In the $(1+1)$ ES algorithm, the parameter $\sigma$ plays a crucial role in the search process. Ideally, $\sigma$ should not remain constant; instead, it should be large in the initial iterations to ensure broad exploration of the solution space, and then decrease over time to refine the solution. Therefore, determining the initial value of $\sigma$ and defining its adjustment over iterations is essential. To address this, Rechenberg proposed an improved version of the $(1+1)$ ES known as adaptive $(1+1)$ ES, which is described in the following section.

### 10.5.1 Adaptive (1 + 1) ES

Adaptive $(1+1)$ ES enhances the search strategy by dynamically adjusting $\sigma$ so that its value changes automatically over time. This method is based on the theory that 20% of the mutations should be successful (i.e., they should improve the quality of the solution) to consider $\sigma$ as appropriate. If the improvement rate exceeds 20%, the mutations are too small, indicating that $\sigma$ is too small, resulting in minor improvements and prolonged convergence time. Conversely, if the improvement rate is below 20%, the mutations are too large, meaning $\sigma$ is too large, leading to significant but infrequent improvements and also extended convergence time.

Rechenberg's 1/5 success rule addresses this by suggesting that if the success rate of mutations is less than 1/5, the standard deviation $\sigma$ should be decreased. If the success rate is higher than 1/5, $\sigma$ should be increased.

Using this rule, we can determine when to adjust the standard deviation $\sigma$. However, the next question is by how much $\sigma$ should be modified. Schwefel proposed a factor for adjusting $\sigma$ according to the 1/5 rule. This factor, a constant $c$, is suggested to be 0.817. Thus, the value of $\sigma$ can be adapted as follows:

$$\sigma = c\sigma. \tag{10.7}$$

$$\sigma = \frac{\sigma}{c}. \tag{10.8}$$

The expression provided in Eq. 10.7 facilitates the decrease of the sigma value, while Eq. 10.8 allows for its increase. To determine the success rate of mutations, it is essential to establish a range of analysis to evaluate the effectiveness of mutations. The goal is to set a time window $w$ during which sigma remains constant, allowing for a reliable measurement of mutation success. This analysis range should be large enough to gather sufficient data on mutation success but not so large that it impedes the sigma adaptation process. A recommended guideline is to set $w$ based on the Eq. 10.9:

$$w = \min(d, 30). \tag{10.9}$$

where $d$ is the number of dimensions in the optimization problem, with $w$ defined and knowing the number of successful mutations $nm$ within that period, the success rate $\varphi$ can be calculated as defined in Eq. 10.10.

$$\varphi = \frac{nm}{w}. \tag{10.10}$$

Therefore, the sigma adaptation follows this procedure described in Eq. 10.11.

$$\sigma = \begin{Bmatrix} c\sigma, & \varphi < \frac{1}{5} \\ \frac{\sigma}{c}, & \varphi > \frac{1}{5} \end{Bmatrix}. \tag{10.11}$$

By incorporating this adaptive mechanism, the algorithm can dynamically adjust $\sigma$ to balance exploration and exploitation, potentially leading to faster and more accurate

## 10.5 Variants of Evolutionary Strategies

convergence to the global optimum. The general procedure for adaptive (1 + 1) ES is summarized in Algorithm 10.2.

---

**Algorithm 10.2**
The general procedure of the adaptive (1+1) ES.

**Step 1:** Parameter configuration
$$d, lb, ub, k \leftarrow 0, k_{max}, \sigma^2, w, c$$

**Step 2:** Particle initialization
$$x_p^k \leftarrow r \cdot (ub - lb) + lb$$

**Step 3:** Evaluation of the initial particle in the objective function
$$fx_p^k \leftarrow f(x_p^k)$$

**Step 4:** Generate mutation vector
$$\rho \leftarrow N(0, \sigma^2)$$

**Step 5:** Mutation
$$x_o^k \leftarrow x_p^k + \rho$$

**Step 6:** Evaluation of the offspring in the objective function
$$fx_o^k \leftarrow f(x_o^k)$$

**Step 7:** Selection
$$\text{if } fx_o^k < fx_p^k$$
$$x_p^k \leftarrow x_o^k$$
$$fx_p^k \leftarrow fx_o^k$$

**Step 8:** Sigma adaptation
$$\varphi = \frac{nm}{w}$$
$$\text{if } \varphi < \frac{1}{5}$$
$$\varphi \leftarrow c\sigma$$
$$\text{else if } \varphi > \frac{1}{5}$$
$$\varphi \leftarrow \frac{\sigma}{c}$$

**Step 9:** Verify the stop condition
$$\text{if } k = k_{max}$$
$$\text{Best solution} \leftarrow x_p^k$$
End the search process
else
   Go to Step 4
end

---

The adaptive $(1+1)$ ES algorithm has been implemented in the MATLAB environment to illustrate the search procedure, as demonstrated in Code 10.2. Similar to Code 10.1, the Rastringin function is utilized, considering two dimensions, 500 iterations, and an initial sigma $\sigma = 2$. Additionally, a factor $c = 0.817$ and a time window $w = 20$ have been included.

```matlab
% Code 10.2
% Adaptive (1+1) ES algorithm

% Clear memory and close windows
clear all
close all
% Parameter configuration
    d = 2;                  % Number of dimensions
    lb = [-5.12 -5.12];% Lower bound
    ub = [5.12 5.12];   % Upper bound
    k = 0;                  % Iteration counter
    kmax = 500;             % Maximum number of iterations
    sigma=2;                % Standard deviation
    sigma2= sigma^2;    % Variance
    w = 20;                 % Window
    c = 0.817;              % Constant factor
    phi = 0;                % Mutation rate
    nm = 0;                 % Number of successful mutations
    cont=0;                 % Counter
% Optimization problem (minimization), objective function definition
    f = @(x) 10*d + sum(x.^2 - 10*cos(2*pi.*x));
% Particle initiañization
    Xp = rand(1,d).*(ub-lb)+lb;
% Evaluation of the initial particle in the objective function
    fXp = f(Xp);
% Determine the search space
    xAxis=linspace(min(lb),max(ub),85);
    yAxis=xAxis;
    zAxis=[];
    for i = 1:length(xAxis)
        for j = 1:length(yAxis)
            zAxis(i,j) = f([xAxis(i) yAxis(j)]);
        end
```

## 10.5 Variants of Evolutionary Strategies

```
          end
       [yAxis,xAxis] = meshgrid(xAxis,yAxis);
    % Iterative process
    while k<kmax
        % Generate mutation vector
           rho = normrnd(0,sigma,1,d);
        % Mutation
           Xo = Xp + rho;
        % Verify the limits of the search space
           for j=1:d
               if Xo(j) < lb(j)
                   Xo(j) = rand*(ub(j)-lb(j))+lb(j);
               elseif Xo(j) > ub(j)
                   Xo(j) = rand*(ub(j)-lb(j))+lb(j);
               end
           end
        % Evaluate the offspring in the objective function
           fXo = f(Xo);
        % Draw the search space
           figure(1);
           surf(xAxis,yAxis,zAxis)
           hold on
        % Draw particles Xp y Xo

plot3(Xp(1),Xp(2),fXp,'o','MarkerFaceColor','m','MarkerSize',10)

plot3(Xo(1),Xo(2),fXo,'o','MarkerFaceColor','y','MarkerSize',10)
           pause(0.01)
           hold off
        % Draw the contour contorno of the search space
           figure(2)
           contour(xAxis,yAxis,zAxis,20)
           hold on
        % Draw Xp y Xo in the contour
           plot(Xp(1),Xp(2),'o','MarkerFaceColor','m');
           plot(Xo(1),Xo(2),'o','MarkerFaceColor','y');
           pause(0.01)
           hold off
        % Selection
          if fXo < fXp
              Xp = Xo;
              fXp = fXo;
              nm = nm + 1;
          end
        % Sigma adaptation
          if k > 30
                cont = cont + 1;
                if cont == w
                    phi = nm/w;
                    if phi < 1/5
                        sigma = c*sigma;
                    elseif phi > 1/5
                        sigma = sigma/c;
                    end
                    cont = 0;
                    nm = 0;
                end
          end
```

```
        k=k+1;
    end
    % Show results
        display(['Best solution: x1= ',num2str(Xp(1)),', x2= 
    ',num2str(Xp(2))])
        display([' f(X)= ',num2str(fXp)])
```

This MATLAB code demonstrates the adaptive (1 + 1) ES algorithm, showing how the dynamic adjustment of $\sigma$ can improve the search process, balancing exploration and exploitation effectively.

**Performance Comparison of (1 + 1) ES and Adaptive (1 + 1) ES**

After running Code 10.2, it is clear that the adaptive (1 + 1) ES algorithm finds a solution closer to the expected optimal than the standard (1 + 1) ES algorithm. The adaptive method achieves higher precision and reaches the solution faster. However, the adaptive (1 + 1) ES does not always reach the global optimum due to the complexity of the Rastringin function. The adaptive method sometimes gets stuck in local optima due to the randomness and the gradual reduction of the standard deviation. Over time, as sigma adapts, the resulting smaller mutations make it harder to escape local optima, limiting further exploration of potential solutions. Nonetheless, this adaptation allows for the refinement of potential solutions that have already been discovered.

To evaluate the performance of the (1 + 1) ES against the adaptive (1 + 1) ES, both algorithms were executed 100 times using the Rastringin function in 30 dimensions. Each execution consisted of 500 iterations with a standard deviation set to 2. For the adaptive method, a factor $c = 0.817$ and a time window $w = 30$ were used. The average results obtained by both algorithms in each iteration are depicted in Fig. 10.5.

From Fig. 10.5, it can be seen that the adaptive (1 + 1) ES shows a faster convergence rate compared to the standard (1 + 1) ES, indicating its efficiency in reaching near-optimal solutions more quickly. Furthermore, the adaptive method achieves higher precision in finding solutions, as reflected in the lower average fitness values over the iterations. Despite the improved performance, the adaptive (1 + 1) ES occasionally stagnates in local optima due to the diminishing sigma, which limits its ability to explore new areas in the solution space. The overall performance indicates that the adaptive (1 + 1)

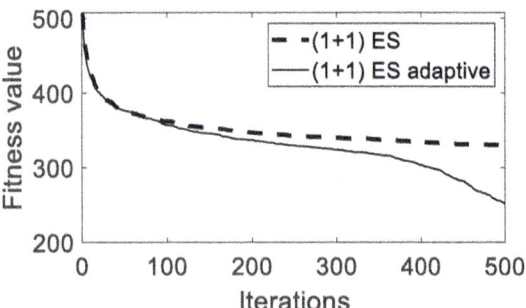

**Fig. 10.5** Performance comparison of algorithm (1 + 1) ES and adaptive (1 + 1) ES

## 10.5 Variants of Evolutionary Strategies

**Fig. 10.6** Salomon function

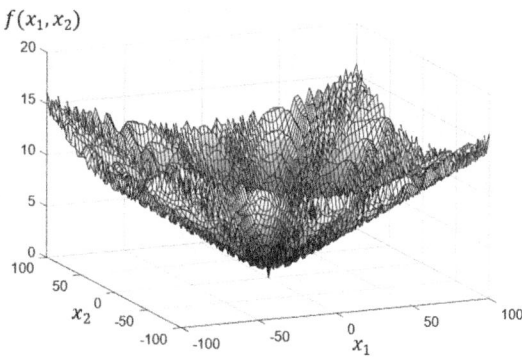

ES outperforms the standard (1 + 1) ES in terms of both speed and accuracy, although it is not foolproof in avoiding local optima.

The stagnation in local optima of the adaptive (1 + 1) ES is because the success of this method depends on several factors, such as the number of dimensions of the problem, the surface characteristics of the objective function, the proper selection of the initial sigma value, and the determination of the time window. To verify that the performance success is not always consistent, both methods were executed 100 times using the Salomon function in 30 dimensions, with the same parameters as before, where the Salomon function is defined in Eq. 10.12.

$$f(x) = 1 - \cos\left(2\pi \sqrt{\sum_{i=1}^{d} x_i^2}\right) + 0.1\sqrt{\sum_{i=0}^{d} x_i^2} .  \quad (10.12)$$

$$x_i \in [-100, 100]; i \in \{1, ..., d\}$$

The Salomon function is illustrated in Fig. 10.6. It has a global optimum $f(x^*) = 0$ at $x^* = \{0, ..., 0\}$ and multiple local optima on the surface.

The results, depicted in Fig. 10.7, show how the adaptive (1 + 1) ES method tends to stagnate in sub-optimal solutions, while the standard (1 + 1) ES manages to get closer to the optimal solution. The adaptive method can easily get stuck in a sub-optimal solution when the objective function is highly complex, and the initial sigma value and the time window size are inappropriate.

While the 1/5 rule does not always yield optimal results and requires careful tuning of parameters, it provides a functional and general implementation guide for the Evolutionary Strategies algorithm. The adaptive (1 + 1) ES algorithm, with its dynamic adjustment of sigma, shows significant improvement over the standard (1 + 1) ES in many cases. However, its effectiveness can vary depending on the complexity of the objective function and the appropriateness of the parameters used.

**Fig. 10.7** Performance comparison of the (1 + 1) ES and the adaptive (1 + 1) ES algorithm using the Salomon function

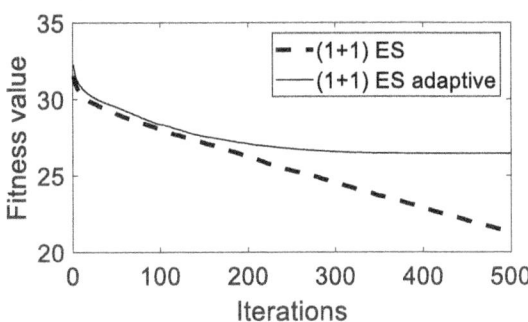

### 10.5.2 The (µ + 1) ES

Following the development of the (1 + 1) ES algorithm, the (µ + 1) ES method was introduced. This variant leverages a population of individuals rather than a single particle, aiming to enhance the search process by facilitating information exchange among particles. The primary advantage of using a population is the acceleration of the search process, achieved through shared information between individuals.

The (µ + 1) ES algorithm introduces several parents, denoted as µ, in each generation. Each parent $x_{pi}$ is associated with a vector $\sigma_{pi}$ that determines the magnitude of its movements in each dimension. The algorithm operates as follows:

1. **Parent Selection**: Two parents, $x_{pa}$ and $x_{pb}$, are randomly chosen from the population.
2. **Recombination**: These selected parents undergo recombination to produce a single offspring $x_o$.
3. **Mutation**: The offspring $x_o$ is then mutated.
4. **Selection**: From the pool of µ parents and the single offspring, the best µ individuals are selected to form the new population of parents for the next generation.

The flowchart for the (µ + 1) ES algorithm is depicted in Fig. 10.8. Comparing this flowchart with Fig. 10.3, which represents the (1 + 1) ES method, reveals two significant differences:

1. **Population-based Approach**: Unlike the (1 + 1) ES, which uses a single individual, the (µ + 1) ES employs a population of µ individuals.
2. **Recombination Operator**: The (µ + 1) ES incorporates a recombination step, which is absent in the (1 + 1) ES. This operator combines genetic material from two parents, potentially leading to more diverse and robust offspring.

## 10.5 Variants of Evolutionary Strategies

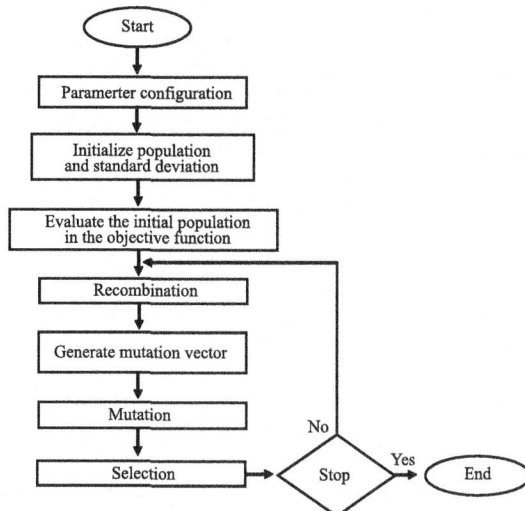

**Fig. 10.8** Flowchart of the (μ + 1) ES algorithm

The (μ + 1) ES algorithm introduces a population-based approach and a recombination operator, which represent significant advancements over the (1 + 1) ES. These modifications not only enhance the search efficiency but also improve the algorithm's ability to explore the solution space comprehensively, increasing the likelihood of finding optimal or near-optimal solutions.

In the (μ + 1) ES algorithm, recombination is a crucial mechanism that facilitates the exchange of information between parents. This process aims to create an offspring that inherits a blend of characteristics from its parents. Specifically, the recombination method used in (μ + 1) ES not only combines the positional information of the parents to generate the child but also merges their standard deviation vectors to produce the offspring's $\sigma_o$ vector.

There are various recombination methods, including:

1. Discrete Sexual Recombination
2. Intermediate Recombination
3. Global or Panmitic Recombination
4. Global Discrete Recombination
5. Global Intermediate Recombination.

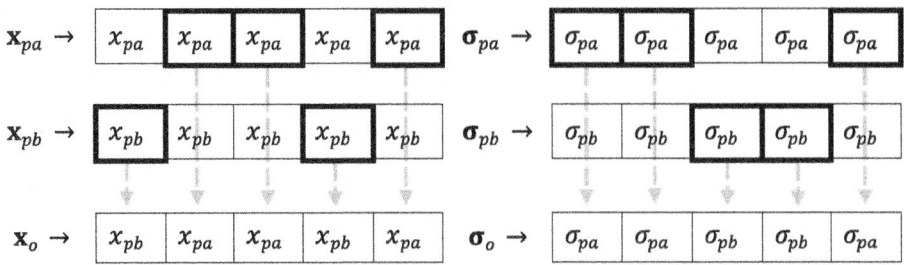

**Fig. 10.9** Example of discrete sexual recombination where each element of $x_o$ is randomly selected from the parents $x_{pa}$ and $x_{pb}$

**Discrete Sexual Recombination**

In this method, the offspring $x_o$ is formed by randomly selecting elements from each parent $x_{pa}$ and $x_{pb}$. The standard deviation vector $\sigma_o$ of the offspring is similarly constructed by randomly choosing elements from the parents' standard deviation vectors $\sigma_{pa}$ and $\sigma_{pb}$. This process is illustrated in Fig. 10.9.

**Intermediate Sexual Recombination**

This method involves averaging the information from both parents to generate the offspring. For example, each element of the offspring is the average of the corresponding elements from both parents. Figure 10.10 illustrates this process.

**Global or Panmictic Recombination**

Here, the offspring is constructed by considering all parents in the population as potential contributors. Unlike discrete sexual recombination, which uses only two parents, global recombination allows any parent to contribute to the offspring's formation.

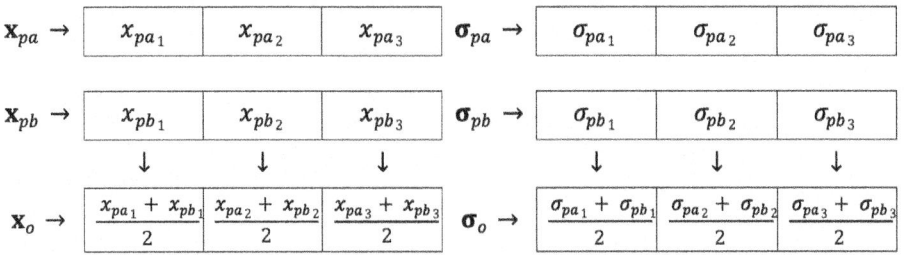

**Fig. 10.10** Example of intermediate sexual recombination where each element of $x_o$ is calculated with the mean of $x_{pa}$ and $x_{pb}$

## 10.5 Variants of Evolutionary Strategies

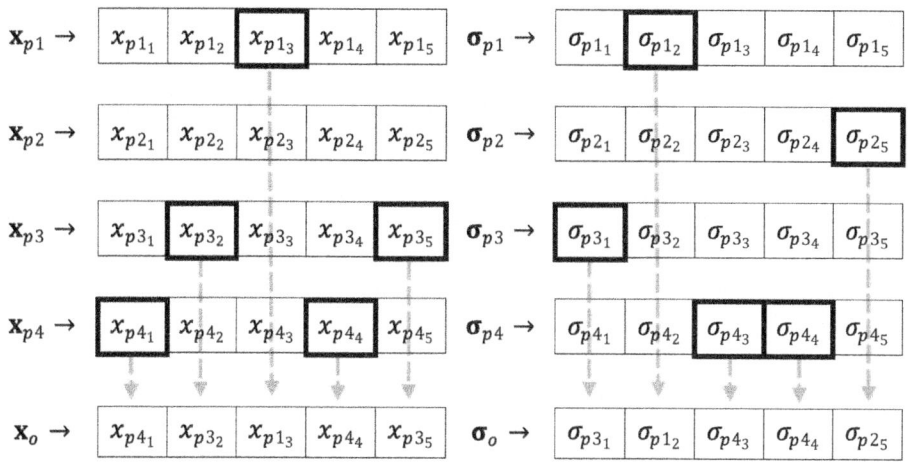

**Fig. 10.11** Example of global discrete recombination considering a population of four individuals, where each element of $x_o$ is randomly chosen from the population

**Global Discrete Recombination**

This method combines discrete sexual recombination with global recombination. Each element of the offspring is formed from a randomly selected parent from the entire population. Figure 10.11 provides an example of this type of recombination.

**Global Intermediate Recombination**

In this method, each element of the offspring is determined by averaging the information from two parents randomly chosen from the entire population. Unlike sexual intermediate recombination, where two parents are chosen once and used for all elements, global intermediate recombination selects a new pair of parents for each element. This method is illustrated in Fig. 10.12.

These methods enhance the search process by diversifying the genetic material available to the offspring, improving the algorithm's ability to explore and exploit the search space effectively. Other recombination methods have been proposed, but they are beyond the scope of this chapter.

**Differences between (1 + 1) ES and (M + 1) ES**

The $(\mu + 1)$ ES algorithm introduces several key differences compared to the $(1 + 1)$ ES:

1. **Population-Based Approach**: Unlike the $(1 + 1)$ ES, which operates with a single parent, the $(\mu + 1)$ ES uses a population of $\mu$ parents. This allows for information exchange between individuals, aiming to accelerate the search process.

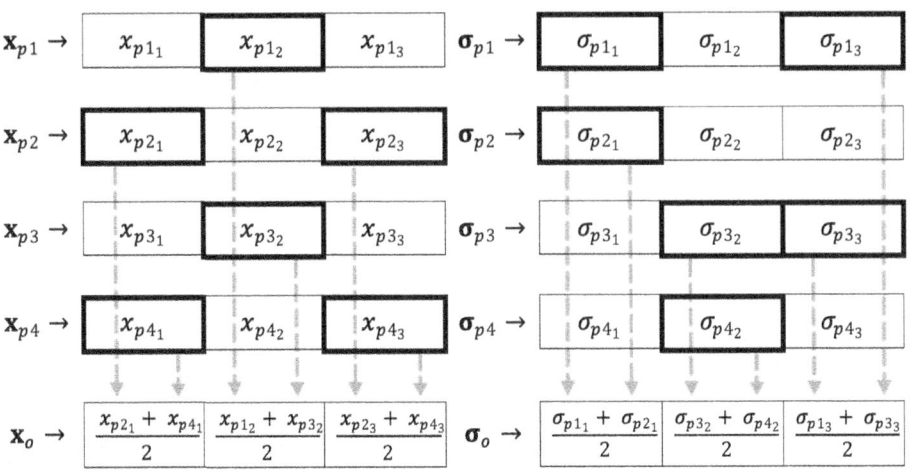

**Fig. 10.12** An example of global intermediate recombination considering a population of four individuals, where each element of $x_o$ is determined with the average of two randomly selected parents

2. **Mutation Mechanism**: In the ($\mu + 1$) ES, it is the offspring, not the parent, that undergoes mutation. Additionally, the variance of each element of the offspring $x_o$ is different.
3. **Non-Isometric Mutations**: The ($\mu + 1$) ES mutations are non-isometric, meaning that each dimension can have a different variance. The mutation for each child is defined by Eq. 10.13.

$$x_o = x_o + \rho. \tag{10.13}$$

where $\rho = \{\rho_1, ..., \rho_d\}$ is a vector of normally distributed random values, with each element given by Eq. 10.14.

$$\rho_j = N(0, \sum)$$
$$j = 1, 2, ..., d. \tag{10.14}$$

4. **Multivariate Normal Distribution**: The mutation operation uses a multivariate normal distribution with zero mean and a covariance matrix $\Sigma$ defined by Eq. 10.15.

## 10.5 Variants of Evolutionary Strategies

$$\Sigma = \begin{bmatrix} \sigma_{o1}^2 & 0 & \cdots & 0 \\ 0 & \sigma_{o2}^2 & \cdots & 0 \\ \vdots & \vdots & \ddots & \vdots \\ 0 & 0 & \cdots & \sigma_{od}^2 \end{bmatrix}. \tag{10.15}$$

Here, $\Sigma$ is a $d \times d$ symmetric matrix, with the variances for each element of $\sigma_o$ along the main diagonal.

Figure 10.13 illustrates a normal non-isometric bivariate distribution.

On the other hand, Fig. 10.14 shows potential positions of a mutated particle around $x_o$.

Figure 10.15 compares the mutation distributions of the $(1 + 1)$ ES and $(\mu + 1)$ ES methods.

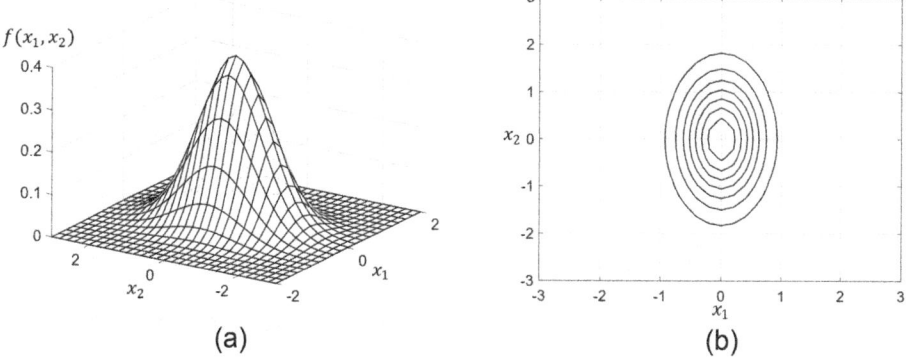

**Fig. 10.13** a Normal non-isometric bivariate distribution and b its top view

**Fig. 10.14** Distribution of mutated particles in the algorithm $(\mu + 1)$ ES

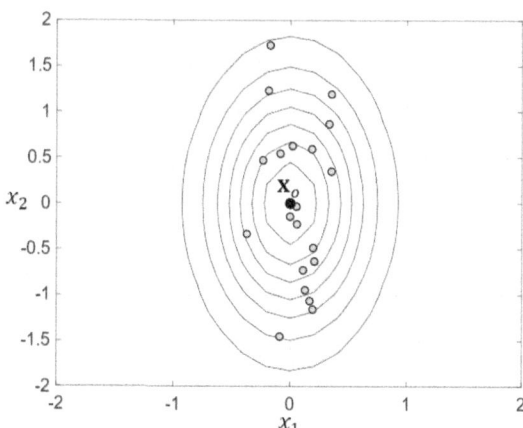

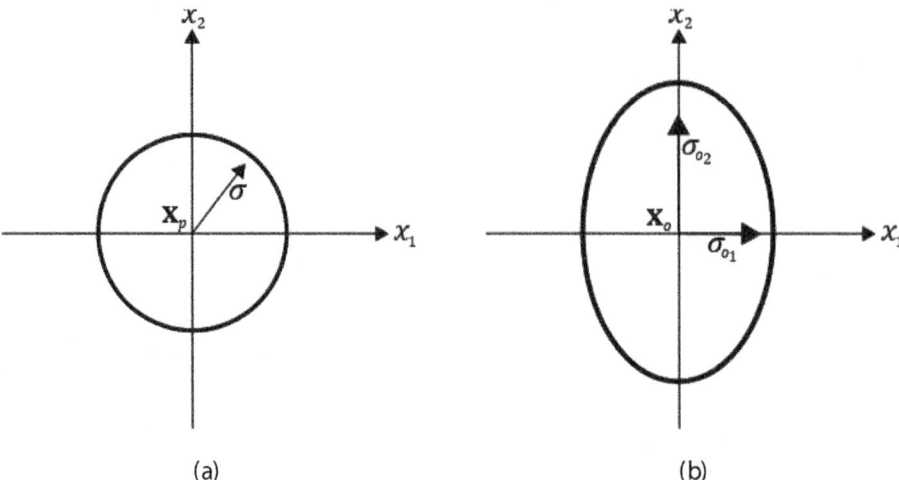

**Fig. 10.15** a Mutation distribution of the (1 + 1) ES and b the (μ + 1) ES

**General procedure of the $(\mu + 1)$ ES**

The $(\mu + 1)$ ES method's general procedure, utilizing discrete sexual recombination, is detailed in Algorithm 10.3.

## 10.5 Variants of Evolutionary Strategies

---

**Algorithm 10.3**
The general procedure of the ($\mu$+1) ES method.

**Step 1:** Parameter configuration
$$\mu, d, lb, ub, k \leftarrow 0, k_{max}$$

**Step 2:** Particle and standard deviation initialization
$$for\ i = 1; i \leq \mu; i++$$
$$x_{pi}^k \leftarrow r \cdot (ub - lb) + lb$$
$$\sigma_{pi}^k \leftarrow rand(); \sigma_{pi}^k \in \mathbb{R}^d$$
$$end$$

**Step 3:** Evaluation of the initial particles in the objective function
$$for\ i = 1; i \leq \mu; i++$$
$$fx_{pi}^k \leftarrow f(x_{pi}^k)$$
$$end$$

**Step 4:** Select parents
$$(x_{pa}^k, \sigma_{pa}^k) \leftarrow rand(x_p^k, \sigma_p^k)$$
$$(x_{pb}^k, \sigma_{pb}^k) \leftarrow rand(x_p^k, \sigma_p^k)$$

**Step 5:** Recombination
$$x_o^k \leftarrow recombination(x_{pa}^k, x_{pb}^k)$$
$$\sigma_o^k \leftarrow recombination(\sigma_{pa}^k, \sigma_{pb}^k)$$

**Step 6:** Generate mutation vector
$$\Sigma \leftarrow diag\left(\left(\sigma_{o_1}^k\right)^2, \ldots, \left(\sigma_{o_d}^k\right)^2\right) \in \mathbb{R}^{d \times d}$$
$$\rho \leftarrow N(0, \Sigma)$$

**Step 7:** Mutation
$$x_o^k \leftarrow x_p^k + \rho$$

**Step 8:** Evaluation of the offspring in the objective function
$$fx_o^k \leftarrow f(x_o^k)$$

**Step 9:** Selection
$$x_p^k \leftarrow best(\{x_p^k \cup x_o^k\}, \mu)$$

**Step 10:** Verify the stop condition
$$if\ k = k_{max}$$
$$Best\ solution \leftarrow x_p^k$$
$$End\ the\ search\ process$$
$$else$$
$$k \leftarrow k+1$$
$$Go\ to\ Step\ 4$$
$$end$$

Code 10.3 implements this in MATLAB for the Rastringin function in two dimensions, with a population size of 3 and 500 iterations. This implementation reveals that while the ($\mu + 1$) ES often approximates the global minimum, it may not consistently achieve it.

The ($\mu + 1$) ES does not employ the adaptive strategy based on the 1/5 rule, unlike the adaptive (1 + 1) ES. Instead, it relies on the recombination of the standard deviation vectors associated with each individual. This recombination process modifies the variances in each direction and adjusts the size of the particles' movements. Over time, the variances tend to converge towards a single value due to the continuous recombination.

Executing Code 10.3 in MATLAB, it is observed that the ($\mu + 1$) ES method often approximates the global minimum, although it may not consistently achieve the expected result. To enhance the ($\mu + 1$) ES algorithm, various similar variants have been developed and will be explored in subsequent sections.

```matlab
% Code 10.3
% (μ+1) ES algorithm

% Clear memory and close windows
clear all
close all
% Parameter configuration
    mu = 3;                 % Number of particles
    d = 2;                  % Dimensions
    lb = [-5.12 -5.12];     % Lower bound
    ub = [5.12 5.12];       % Upper bound
    k = 0;                  % Iteration counter
    kmax = 500;             % Maximum number of iterations
% Optimization problem (minimization), objective function definition
    f = @(x) 10*d + sum(x.^2 - 10*cos(2*pi.*x));
% Initialize particles and standard deviations
    for i = 1:mu
        Xp(i,:) = rand(1,d).*(ub-lb)+lb;
        sigma_p(i,:) = rand(1,d)*2;
    end
% Evaluate initial particles in the objective function
    for i = 1:mu
        fXp(i,1) = f(Xp(i,:));
```

## 10.5 Variants of Evolutionary Strategies

```matlab
        end
    % Definition of the search space
        xAxis=linspace(min(lb),max(ub),85);
        yAxis=xAxis;
        zAxis=[];
        for i = 1:length(xAxis)
            for j = 1:length(yAxis)
                zAxis(i,j) = f([xAxis(i) yAxis(j)]);
            end
        end
        [yAxis,xAxis] = meshgrid(xAxis,yAxis);
    % Iterative process
    while k<kmax
        % Parents selection
            % Parent 1
            a = randi(mu);
            Xpa   = Xp(a,:);
            sigma_pa = sigma_p(a,:);
            % Parent 2
            b = randi(mu);
            Xpb   = Xp(b,:);
            sigma_pb = sigma_p(b,:);
        % Recombination
            for i=1:d
                if rand<0.5
                    Xo(1,i) = Xpa(i);
                    sigma_o(1,i) = sigma_pa(i);
                else
                    Xo(1,i) = Xpb(i);
                    sigma_o(1,i) = sigma_pb(i);
                end
            end
        % Generate mutation vector
            rho = normrnd(0,sigma_o);
        % Mutation
            Xo = Xo + rho;
        % Verify bounds
            for j=1:d
                if Xo(j) < lb(j)
                    Xo(j) = rand*(ub(j)-lb(j))+lb(j);
                elseif Xo(j) > ub(j)
                    Xo(j) = rand*(ub(j)-lb(j))+lb(j);
                end
            end
        % Evaluate the offspring in the objective function
            fXo = f(Xo);
        % Draw the search space
            figure(1);
            surf(xAxis,yAxis,zAxis)
            hold on
        % Draw particles Xp y Xo
plot3(Xp(:,1),Xp(:,2),fXp,'o','MarkerFaceColor','m','MarkerSize',10)

plot3(Xo(1),Xo(2),fXo,'o','MarkerFaceColor','y','MarkerSize',10)
            pause(0.01)
            hold off
        % Draw contour
```

```
        figure(2)
        contour(xAxis,yAxis,zAxis,20)
        hold on
    % Draw Xp y Xo in contour
        plot(Xp(:,1),Xp(:,2),'o','MarkerFaceColor','m');
        plot(Xo(1),Xo(2),'o','MarkerFaceColor','y');
        pause(0.01)
        hold off
    % Select the best mu individuals
        Xpo = [Xp;Xo];
        sigma_po = [sigma_p;sigma_o];
        fXpo = [fXp;fXo];
        [fXpo, ind] = sort(fXpo);
        Xpo = Xpo(ind,:);
        sigma_po = sigma_po(ind,:);
        Xp = Xpo(1:mu,:);
        sigma_p = sigma_po(1:mu,:);
        fXp = fXpo(1:mu);
    k=k+1;
end
% Show results
    display(['Best solution: x1= ',num2str(Xp(1,1)),', x2= ',num2str(Xp(1,2))])
    display(['  f(X)= ',num2str(fXp(1))])
```

### 10.5.3 The $(\mu + \lambda)$ ES

Similar to the $(\mu + 1)$ ES algorithm, the $(\mu + \lambda)$ ES method starts with an initial population of $\mu$ parents. However, unlike $(\mu + 1)$ ES, which generates a single offspring per iteration, $(\mu + \lambda)$ ES produces $\lambda$ offspring through recombination between parents in each iteration. After generating the offspring, the combined population of $\mu$ parents and $\lambda$ offspring, totaling $\mu + \lambda$ individuals, is evaluated. The best $\mu$ individuals from this combined set are then selected to form the new parent population for the next generation.

The general procedure of the $(\mu + \lambda)$ ES algorithm is detailed in Algorithm 10.4, which employs discrete sexual recombination.

## 10.5 Variants of Evolutionary Strategies

**Algorithm 10.4**
The general procedure of the $(\mu + \lambda)$ ES method.

Step 1: Parameter configuration
$$\mu, \lambda, d, lb, ub, k \leftarrow 0, k_{max}$$

Step 2: Particle and standard deviation initialization
$$\text{for } i = 1; i \leq \mu; i++$$
$$x_{pi}^k \leftarrow r \cdot (ub - lb) + lb$$
$$\sigma_{pi}^k \leftarrow rand(); \sigma_{pi}^k \in \mathbb{R}^d$$
$$\text{end}$$

Step 3: Evaluation of the initial particles in the objective function
$$\text{for } i = 1; i \leq \mu; i++$$
$$fx_{pi}^k \leftarrow f(x_{pi}^k)$$
$$\text{end}$$

Step 4: Select parents
$$(x_{pa}^k, \sigma_{pa}^k) \leftarrow rand(x_p^k, \sigma_p^k)$$
$$(x_{pb}^k, \sigma_{pb}^k) \leftarrow rand(x_p^k, \sigma_p^k)$$

Step 5: Recombination
$$x_{on}^k \leftarrow recombination(x_{pa}^k, x_{pb}^k)$$
$$\sigma_{on}^k \leftarrow recombination(\sigma_{pa}^k, \sigma_{pb}^k)$$

Step 6: Generate mutation vector
$$\Sigma \leftarrow diag\left(\left(\sigma_{on_1}^k\right)^2, \ldots, \left(\sigma_{on_d}^k\right)^2\right) \in \mathbb{R}^{d \times d}$$
$$\rho \leftarrow N(0, \Sigma)$$

Step 7: Mutation
$$x_{on}^k \leftarrow x_{on}^k + \rho$$

Step 8: Evaluation of the offspring in the objective function
$$fx_{on}^k \leftarrow f(x_{on}^k)$$

Step 9: Verification of the number of generated particles
$$\text{if } n == \lambda$$
$$\quad \text{Go to step 10}$$
$$\text{else}$$
$$\quad n \leftarrow n+1$$
$$\quad \text{Go to Step 4}$$
$$\text{end}$$

Step 10: Selection
$$x_p^k \leftarrow best(\{x_p^k \cup x_o^k\}, \mu)$$

> **Step 11:** Verify the stop condition
> $$\text{if } k = k_{max}$$
> $$\text{Best solution} \leftarrow x_p^k$$
> $$\text{End the search process}$$
> $$\text{else}$$
> $$k \leftarrow k+1$$
> $$\text{Go to Step 4}$$
> $$\text{end}$$

Algorithm 10.4 illustrates that in each generation, two parents are randomly selected and recombined to create an offspring, which is subsequently mutated. This process is repeated until λ offspring are generated. Following this, the selection phase involves combining the λ offspring with the μ parents to form a set of μ + λ individuals. From this set, the best μ individuals are chosen to become the parents of the next generation. This algorithm is implemented in MATLAB as shown in Code 10.4, using the Rastrigin function in two dimensions, with a population size of μ = 3, an offspring size of λ = 6, and 500 iterations.

```
% Code 10.4
% (μ+λ) ES method

% Clear memory and close windows
clear all
close all
% Parameter configuration
    mu = 3;              % Number of particles
    lambda = 6;          % Number of children
    d = 2;               % Dimension
    lb = [-5.12 -5.12];  % Lower bound
    ub = [5.12 5.12];    % Upper bound
    k = 0;               % Iteration counter
    kmax = 500;          % Maximum number of iterations
% Optimization problem (minimization), objective function definition
    f = @(x) 10*d + sum(x.^2 - 10*cos(2*pi.*x));
% Initialize particles and standard deviations
    for i = 1:mu
        Xp(i,:) = rand(1,d).*(ub-lb)+lb;
        sigma_p(i,:) = rand(1,d)*2;
    end
% Evaluate initial particles in the objective function
    for i = 1:mu
        fXp(i,1) = f(Xp(i,:));
    end
% Definition of the search space
    xAxis=linspace(min(lb),max(ub),85);
    yAxis=xAxis;
    zAxis=[];
    for i = 1:length(xAxis)
        for j = 1:length(yAxis)
            zAxis(i,j) = f([xAxis(i) yAxis(j)]);
        end
    end
```

## 10.5 Variants of Evolutionary Strategies

```matlab
        [yAxis,xAxis] = meshgrid(xAxis,yAxis);
% Iterative process
while k<kmax
    for n=1:lambda
        % Parents selection
            % Parent 1
            a = randi(mu);
            Xpa     = Xp(a,:);
            sigma_pa = sigma_p(a,:);
            % Parent 2
            b = randi(mu);
            Xpb     = Xp(b,:);
            sigma_pb = sigma_p(b,:);
        % Recombination
            for i=1:d
                if rand<0.5
                    Xo(n,i) = Xpa(i);
                    sigma_o(n,i) = sigma_pa(i);
                else
                    Xo(n,i) = Xpb(i);
                    sigma_o(n,i) = sigma_pb(i);
                end
            end
        % Generate mutation vector
            rho = normrnd(0,sigma_o(n,:));
        % Mutation
            Xo(n,:) = Xo(n,:) + rho;
        % Verify bounds
            for j=1:d
                if Xo(n,j) < lb(j)
                    Xo(n,j) = rand*(ub(j)-lb(j))+lb(j);
                elseif Xo(n,j) > ub(j)
                    Xo(n,j) = rand*(ub(j)-lb(j))+lb(j);
                end
            end
        % Evaluate the offspring in the objective function
            fXo(n,1) = f(Xo(n,:));
    end
    % Draw the search space
        figure(1);
        surf(xAxis,yAxis,zAxis)
        hold on
    % Draw particles Xp y Xo
plot3(Xp(:,1),Xp(:,2),fXp,'o','MarkerFaceColor','m','MarkerSize',10)

plot3(Xo(:,1),Xo(:,2),fXo,'o','MarkerFaceColor','y','MarkerSize',10)
        pause(0.1)
        hold off
    % Draw contour
        figure(2)
        contour(xAxis,yAxis,zAxis,20)
        hold on
    % Draw Xp y Xo in contour
        plot(Xp(:,1),Xp(:,2),'o','MarkerFaceColor','m');
        plot(Xo(:,1),Xo(:,2),'o','MarkerFaceColor','y');
        pause(0.1)
        hold off
```

```
        % Select the best mu individuals
            Xpo = [Xp;Xo];
            sigma_po = [sigma_p;sigma_o];
            fXpo = [fXp;fXo];
            [fXpo, ind] = sort(fXpo);
            Xpo = Xpo(ind,:);
            sigma_po = sigma_po(ind,:);
            Xp = Xpo(1:mu,:);
            sigma_p = sigma_po(1:mu,:);
            fXp = fXpo(1:mu);
        k=k+1;
    end
    % Show results
        display(['Best solution: x1= ',num2str(Xp(1,1)),', x2= ',num2str(Xp(1,2))])
        display([' f(X)= ',num2str(fXp(1))])
```

## 10.5.4 The $(\mu, \lambda)$ ES

The $(\mu, \lambda)$ Evolution Strategy (ES) operates similar to the $(\mu + \lambda)$ ES, but with a key distinction in their selection processes. Unlike the $(\mu + \lambda)$ ES, which combines the $\mu$ parents and the $\lambda$ offspring to select the best $\mu$ individuals from the entire population, the $(\mu, \lambda)$ ES does not merge parents with their offspring. Instead, it selects the best $\mu$ individuals solely from the $\lambda$ offspring to form the new parent population. This ensures that each generation of parents survives for only one iteration, imposing that all individuals are replaced in every generation and allowing only the top offspring to continue.

The overall procedure for the $(\mu, \lambda)$ ES is detailed in Algorithm 10.5, which uses discrete sexual recombination.

---

**Algorithm 10.5**
The general procedure of the $(\mu, \lambda)$ ES method

Step 1: Parameter configuration
$$\mu, \lambda, d, lb, ub, k \leftarrow 0, k_{max}$$

Step 2: Particle and standard deviation initialization
$$\text{for } i = 1; i \leq \mu; i++$$
$$x_{pi}^k \leftarrow r \cdot (ub - lb) + lb$$
$$\sigma_{pi}^k \leftarrow rand(); \sigma_{pi}^k \in \mathbb{R}^d$$
$$\text{end}$$

Step 3: Evaluation of the initial particles in the objective function
$$\text{for } i = 1; i \leq \mu; i++$$

## 10.5 Variants of Evolutionary Strategies

$$fx_{pi}^k \leftarrow f(x_{pi}^k)$$
end

**Step 4: Select parents**
$$(x_{pa}^k, \sigma_{pa}^k) \leftarrow rand(x_p^k, \sigma_p^k)$$
$$(x_{pb}^k, \sigma_{pb}^k) \leftarrow rand(x_p^k, \sigma_p^k)$$

**Step 5: Recombination**
$$x_{on}^k \leftarrow recombination(x_{pa}^k, x_{pb}^k)$$
$$\sigma_{on}^k \leftarrow recombination(\sigma_{pa}^k, \sigma_{pb}^k)$$

**Step 6: Generate mutation vector**
$$\Sigma \leftarrow diag\left(\left(\sigma_{on_1}^k\right)^2, ..., \left(\sigma_{on_d}^k\right)^2\right) \in \mathbb{R}^{d \times d}$$
$$\rho \leftarrow N(0, \Sigma)$$

**Step 7: Mutation**
$$x_{on}^k \leftarrow x_{on}^k + \rho$$

**Step 8: Evaluation of the offspring in the objective function**
$$fx_{on}^k \leftarrow f(x_{on}^k)$$

**Step 9: Verification of the number of generated particles**
if $n = \lambda$
   Go to step 10
else
   $n \leftarrow n+1$
   Go to Step 4
end

**Step 10: Selection**
$$x_p^k \leftarrow best(x_o^k, \mu)$$

**Step 11: Verify the stop condition**
if $k = k_{max}$
   Best solution $\leftarrow best(x_p^k)$
   End the search process
else
   $k \leftarrow k+1$
   Go to Step 4
end

Code 10.5, implemented in MATLAB, demonstrates this method using the Rastringin function in two dimensions, with a population size $\mu = 3$, an offspring size $\lambda = 6$, and 500 iterations.

```matlab
% Code 10.5
% (μ,λ) ES method

% Clear memory and close windows
clear all
close all
% Parameter configuration
    mu = 3;             % Number of particles
    lambda = 6;         % Number of children
    d = 2;              % Dimensions
    lb = [-5.12 -5.12]; % Lower bound
    ub = [5.12 5.12];   % Upper bound
    k = 0;              % Iteration counter
    kmax = 500;         % Maximum number of iterations
% Optimization problem (minimization), objective function definition
    f = @(x) 10*d + sum(x.^2 - 10*cos(2*pi.*x));
% Initialize particles and standard deviation
    for i = 1:mu
        Xp(i,:) = rand(1,d).*(ub-lb)+lb;
        sigma_p(i,:) = rand(1,d)*2;
    end
% Evaluate initial particles in the objective function
    for i = 1:mu
        fXp(i,1) = f(Xp(i,:));
    end
% Definition of the search space
    xAxis=linspace(min(lb),max(ub),85);
    yAxis=xAxis;
    zAxis=[];
    for i = 1:length(xAxis)
        for j = 1:length(yAxis)
            zAxis(i,j) = f([xAxis(i) yAxis(j)]);
        end
    end
    [yAxis,xAxis] = meshgrid(xAxis,yAxis);
% Iterative process
while k<kmax
    for n=1:lambda
        % Parents selection
            % Parent 1
            a = randi(mu);
            Xpa = Xp(a,:);
            sigma_pa = sigma_p(a,:);
            % Parent 2
            b = randi(mu);
            Xpb = Xp(b,:);
            sigma_pb = sigma_p(b,:);
        % Recombination
```

## 10.5 Variants of Evolutionary Strategies

```
                for i=1:d
                    if rand<0.5
                        Xo(n,i) = Xpa(i);
                        sigma_o(n,i) = sigma_pa(i);
                    else
                        Xo(n,i) = Xpb(i);
                        sigma_o(n,i) = sigma_pb(i);
                    end
                end
            % Generate mutation vector
                rho = normrnd(0,sigma_o(n,:));
            % Mutation
                Xo(n,:) = Xo(n,:) + rho;
            % Verify bounds
                for j=1:d
                    if Xo(n,j) < lb(j)
                        Xo(n,j) = rand*(ub(j)-lb(j))+lb(j);
                    elseif Xo(n,j) > ub(j)
                        Xo(n,j) = rand*(ub(j)-lb(j))+lb(j);
                    end
                end
            % Evaluate offspring in the objective function
                fXo(n,1) = f(Xo(n,:));
        end
        % Draw search space
            figure(1);
            surf(xAxis,yAxis,zAxis)
            hold on
        % Draw particles Xp y Xo

plot3(Xo(:,1),Xo(:,2),fXo,'o','MarkerFaceColor','y','MarkerSize',10)

plot3(Xp(:,1),Xp(:,2),fXp,'o','MarkerFaceColor','m','MarkerSize',10)
            pause(0.01)
            hold off
        % Draw contour
            figure(2)
            contour(xAxis,yAxis,zAxis,20)
            hold on
        % Draw Xp y Xo in contour
            plot(Xo(:,1),Xo(:,2),'o','MarkerFaceColor','y');
            plot(Xp(:,1),Xp(:,2),'o','MarkerFaceColor','m');
            pause(0.01)
            hold off
        % Select the best mu individual
            [fXo, ind] = sort(fXo);
            Xo = Xo(ind,:);
            sigma_o = sigma_o(ind,:);
            Xp = Xo(1:mu,:);
            sigma_p = sigma_o(1:mu,:);
            fXp = fXo(1:mu);
        k=k+1;
end
% Show results
    display(['Best solution: x1= ',num2str(Xp(1,1)),', x2= ',num2str(Xp(1,2))])
    display([' f(X)= ',num2str(fXp(1))])
```

Typically, the ($\mu + \lambda$) ES method outperforms the ($\mu, \lambda$) ES, particularly in discrete problems with constrained and relatively noise-free search spaces. This is because the ($\mu$,

**Fig. 10.16** Performance analysis of the ($\mu + \lambda$) and ($\mu, \lambda$) ES algorithm

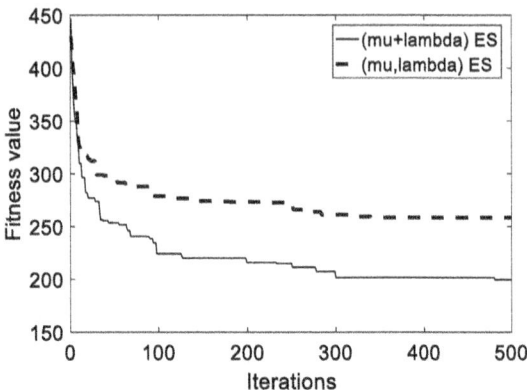

$\lambda$) ES emphasizes exploration more than the ($\mu + \lambda$) ES, as it does not allow an individual to stay in the population for more than one generation. Consequently, it often discards good solutions without refining them. Conversely, the ($\mu, \lambda$) ES performs better in noisy and time-variant objective functions due to its capability to escape local optima, leading to the discovery of better solutions.

To compare the performance of both algorithms, we executed each 100 times using the Rastrigin function in 30 dimensions, with a population size of $\mu = 20$, an offspring size of $\lambda = 30$, and 500 iterations. The average results obtained by both methods in each iteration are illustrated in Fig. 10.16.

As illustrated in Fig. 10.16, the ($\mu + \lambda$) ES algorithm outperforms the ($\mu, \lambda$) ES method. This is because the ($\mu, \lambda$) ES tends to over-explore the search space and discard good solutions prematurely since they are not allowed to survive beyond one generation. Additionally, the bounded nature and moderate noise level of the Rastrigin function also influence the performance.

From this analysis, it can be concluded that the optimal selection of the method depends on the specific problem being optimized. Non-bounded and noisy target functions with multiple local optima require extensive exploration, making the ($\mu, \lambda$) ES more suitable in such scenarios. For other cases, the ($\mu + \lambda$) ES is recommended. However, it is often challenging to precisely determine which method to use for a particular problem due to the typically unknown nature of the search space surface.

### 10.5.5 The ($\mu, \alpha, \lambda, \beta$) ES

The ($\mu, \alpha, \lambda, \beta$) ES method extends the ($\mu + \lambda$) ES and ($\mu, \lambda$) ES algorithms by incorporating two additional parameters, $\alpha$ and $\beta$, alongside the existing $\mu$ parents and $\lambda$ children.

## 10.5 Variants of Evolutionary Strategies

The parameter α controls the number of generations an individual can remain in the population, while β determines the number of parents involved in recombination to generate each child.

If α is set to 1, the (μ, α, λ, β) ES method functions like the (μ, λ) ES, as each individual is only allowed to exist for one generation. Conversely, if α is set to infinity (∞), the algorithm behaves like the (μ + λ) ES, permitting individuals to remain in the population indefinitely as long as they are among the best performers. Thus, adjusting α helps prevent stagnation in local optima by discarding particles that do not improve after a certain number of generations.

On the other hand, the parameter β facilitates more complex recombination strategies, such as global discrete or global intermediate recombination, as discussed in Sect. 10.5.2. With this approach, the offspring's genetic information is derived from more than just two parents, promoting greater diversity within the population.

### 10.5.6 Adaptive (μ + λ) ES and (μ, λ) ES

Up to this point, none of the previously discussed methods, except the adaptive (1 + 1) ES, utilize an adaptive strategy to adjust the standard deviation vector σ. The adaptive (1 + 1) ES employs the 1/5 rule to automatically adjust sigma values based on the success of mutations over a predefined time window. However, this rule is not directly applicable to population methods such as (μ + λ) ES or (μ, λ) ES, where offspring are generated through both mutation and recombination. Therefore, an adaptive strategy is proposed to control the σ values for methods involving recombination in their search strategy.

Implementing a strategy to modify the standard deviations enhances algorithm performance by providing better control over the search process. Initially assigning large σ values facilitates extensive exploration of the search space, while reducing these values in later iterations allows for refinement of the best solutions and greater precision. This approach involves mutating the standard deviation vector alongside the offspring to optimize σ values, thereby balancing exploration and exploitation in each generation. The adaptive strategy involves adjusting the offspring's standard deviation vector $\sigma_o$ before mutating $x_o$, as defined by Eq. 10.16.

$$\sigma_o = \sigma_o e^{(\tau\varphi + \tau'\varphi')}. \tag{10.16}$$

where φ is a constant randomly chosen from a normal distribution $N(0,1)$ and the vector φ' (of dimension d) is composed of random constants from the same distribution. The parameters τ and τ' are constants typically set as:

$$\tau = c_1 \left( \frac{1}{\sqrt{2\sqrt{d}}} \right). \tag{10.17}$$

$$\tau' = c_2\left(\frac{1}{\sqrt{2d}}\right). \tag{10.18}$$

In these equations, $c_1$ and $c_2$ are proportional constants generally set to 1, but can be adjusted for improved performance depending on the optimization problem. The term $\tau\varphi$ broadly influences the mutation of $x_o$, while $\tau'\varphi'$ affects each dimension of $x_o$ specifically.

The adaptive ($\mu + \lambda$) ES procedure is outlined in Algorithm 10.6, which uses discrete sexual recombination.

---

**Algorithm 10.6**
The general procedure of the adaptive ($\mu + \lambda$) ES method

Step 1: Parameter configuration
$$\mu, \lambda, d, lb, ub, k \leftarrow 0, k_{max}, c_1, c_2, \tau, \tau'$$

Step 2: Particle and standard deviation initialization
$$\text{for } i = 1; i \leq \mu; i++$$
$$x_{pi}^k \leftarrow r \cdot (ub - lb) + lb$$
$$\sigma_{pi}^k \leftarrow rand(); \sigma_{pi}^k \in \mathbb{R}^d$$
$$end$$

Step 3: Evaluation of the initial particles in the objective function
$$\text{for } i = 1; i \leq \mu; i++$$
$$fx_{pi}^k \leftarrow f(x_{pi}^k)$$
$$end$$

Step 4: Select parents

## 10.5 Variants of Evolutionary Strategies

$$(x_{pa}^k, \sigma_{pa}^k) \leftarrow rand(x_p^k, \sigma_p^k)$$
$$(x_{pb}^k, \sigma_{pb}^k) \leftarrow rand(x_p^k, \sigma_p^k)$$

**Step 5: Recombination**

$$x_{on}^k \leftarrow recombination(x_{pa}^k, x_{pb}^k)$$
$$\sigma_{on}^k \leftarrow recombination(\sigma_{pa}^k, \sigma_{pb}^k)$$

**Step 6: Iterative process**

$$\varphi \leftarrow N(0,1)$$
$$\varphi' \leftarrow N(0,1); \varphi' \in \mathbb{R}^d$$
$$\sigma_{on}^k \leftarrow \sigma_{on}^k e^{(\tau\varphi + \tau'\varphi')}$$

**Step 7: Generate mutation vector**

$$\Sigma \leftarrow diag\left(\left(\sigma_{on_1}^k\right)^2, \dots, \left(\sigma_{on_d}^k\right)^2\right) \in \mathbb{R}^{d \times d}$$
$$\rho \leftarrow N(0, \Sigma)$$

**Step 8: Mutation**

$$x_{on}^k \leftarrow x_{on}^k + \rho$$

**Step 9: Evaluation of the offspring in the objective function**

$$fx_{on}^k \leftarrow f(x_{on}^k)$$

**Step 10: Verification of the number of generated particles**

    if $n = \lambda$
        Go to step 10
    else
        $n \leftarrow n+1$
        Go to Step 4
    end

**Step 11: Selection**

$$x_p^k \leftarrow best(\{x_p^k \cup x_o^k\}, \mu)$$

**Step 12: Verify the stop condition**

    if $k = k_{max}$
        Best solution $\leftarrow best(x_p^k)$
        End the search process
    else
        $k \leftarrow k+1$
        Go to Step 4
    end

The MATLAB implementation is demonstrated in Code 10.6, where the Rastringin function is used in two dimensions with a population of $\mu = 3$, an offspring of $\lambda = 6$, 500 iterations, and constant parameters $c_1$ and $c_2$ equal to 2.

```matlab
% Code 10.6
% Adaptive (μ+λ) ES
% Clear memory and close windows
clear all
close all
% Parameter configuration
    mu = 3;                 % Number of particles
    lambda = 6;             % Number of children
    d = 2;                  % Dimensions
    lb = [-5.12 -5.12];     % Lower bounds
    ub = [5.12 5.12];       % Upper bounds
    k = 0;                  % Iteration counter
    kmax = 500;             % maximum number of iterations
    c1 = 1;
    c2 = 1;
    tau = c1*(1/(sqrt(2*sqrt(d))));
    tau_prima = c2*(1/(sqrt(2*d)));
% Optimization problem (minimization), objective function definition
    f = @(x) 10*d + sum(x.^2 - 10*cos(2*pi.*x));
% Initializa particles and standard deviation
    for i = 1:mu
        Xp(i,:) = rand(1,d).*(ub-lb)+lb;
        sigma_p(i,:) = rand(1,d)*2;
    end
% Evaluate initial particles in the objective function
    for i = 1:mu
        fXp(i,1) = f(Xp(i,:));
    end
% Definition of the search process
    xAxis=linspace(min(lb),max(ub),85);
    yAxis=xAxis;
    zAxis=[];
    for i = 1:length(xAxis)
        for j = 1:length(yAxis)
            zAxis(i,j) = f([xAxis(i) yAxis(j)]);
        end
    end
    [yAxis,xAxis] = meshgrid(xAxis,yAxis);
% Iterative process
while k<kmax
    for n=1:lambda
        % Parents selection
            % Parent 1
            a = randi(mu);
            Xpa = Xp(a,:);
            sigma_pa = sigma_p(a,:);
            % Parent 2
```

## 10.5 Variants of Evolutionary Strategies

```matlab
            b = randi(mu);
            Xpb   = Xp(b,:);
            sigma_pb = sigma_p(b,:);
         % Recombination
            for i=1:d
                if rand<0.5
                    Xo(n,i) = Xpa(i);
                    sigma_o(n,i) = sigma_pa(i);
                else
                    Xo(n,i) = Xpb(i);
                    sigma_o(n,i) = sigma_pb(i);
                end
            end
         % Adaptative process
            phi = normrnd(0,1);
            phi_prima = normrnd(0,1,1,d);
            sigma_o(n,:) = sigma_o(n,:).*exp(tau*phi +
tau_prima*phi_prima);
         % Generate mutation vector
            rho = normrnd(0,sigma_o(n,:));
         % Mutation
            Xo(n,:) = Xo(n,:) + rho;
         % Verify bounds
            for j=1:d
                if Xo(n,j) < lb(j)
                    Xo(n,j) = rand*(ub(j)-lb(j))+lb(j);
                elseif Xo(n,j) > ub(j)
                    Xo(n,j) = rand*(ub(j)-lb(j))+lb(j);
                end
            end
         % Evaluate offspring in the objective function
            fXo(n,1) = f(Xo(n,:));
    end
    % Draw the search space
        figure(1);
        surf(xAxis,yAxis,zAxis)
        hold on
    % Draw particles Xp y Xo

plot3(Xp(:,1),Xp(:,2),fXp,'o','MarkerFaceColor','m','MarkerSize',10)

plot3(Xo(:,1),Xo(:,2),fXo,'o','MarkerFaceColor','y','MarkerSize',10)
        pause(0.01)
        hold off
    % Draw contour
        figure(2)
        contour(xAxis,yAxis,zAxis,20)
        hold on
    % Draw Xp y Xo in contour
        plot(Xp(:,1),Xp(:,2),'o','MarkerFaceColor','m');
        plot(Xo(:,1),Xo(:,2),'o','MarkerFaceColor','y');
        pause(0.01)
        hold off
    % Select the best mu individual
        Xpo = [Xp;Xo];
        sigma_po = [sigma_p;sigma_o];
        fXpo = [fXp;fXo];
        [fXpo, ind] = sort(fXpo);
```

```
            Xpo = Xpo(ind,:);
            sigma_po = sigma_po(ind,:);
            Xp = Xpo(1:mu,:);
            sigma_p = sigma_po(1:mu,:);
            fXp = fXpo(1:mu);
        k=k+1;
    end
    % Show results
        display(['Best solution: x1= ',num2str(Xp(1,1)),', x2= ',num2str(Xp(1,2))])
        display([' f(X)= ',num2str(fXp(1))])
```

The general procedure of the adaptive ($\mu$, $\lambda$) ES is outlined in Algorithm 10.7, which employs discrete sexual recombination.

---

**Algorithm 10.7**
The general procedure of the adaptive ($\mu$, $\lambda$) ES method

**Step 1: Parameter configuration**
$$\mu, \lambda, d, lb, ub, k \leftarrow 0, k_{\max}, c_1, c_2, \tau, \tau'$$

**Step 2: Particle and standard deviation initialization**
$$\text{for } i = 1; i \leq \mu; i++$$
$$x_{pi}^k \leftarrow r \cdot (ub - lb) + lb$$
$$\sigma_{pi}^k \leftarrow rand(); \sigma_{pi}^k \in \mathbb{R}^d$$
$$end$$

**Step 3: Evaluation of the initial particles in the objective function**
$$\text{for } i = 1; i \leq \mu; i++$$
$$fx_{pi}^k \leftarrow f(x_{pi}^k)$$
$$end$$

**Step 4: Select parents**
$$(x_{pa}^k, \sigma_{pa}^k) \leftarrow rand(x_p^k, \sigma_p^k)$$
$$(x_{pb}^k, \sigma_{pb}^k) \leftarrow rand(x_p^k, \sigma_p^k)$$

**Step 5: Recombination**
$$x_{on}^k \leftarrow recombination(x_{pa}^k, x_{pb}^k)$$
$$\sigma_{on}^k \leftarrow recombination(\sigma_{pa}^k, \sigma_{pb}^k)$$

**Step 6: Iterative process**

## 10.5 Variants of Evolutionary Strategies

$$\varphi \leftarrow N(0,1)$$
$$\varphi' \leftarrow N(0,1); \varphi' \in \mathbb{R}^d$$
$$\sigma_{on}^k \leftarrow \sigma_{on}^k e^{(\tau\varphi + \tau'\varphi')}$$

**Step 7: Generate mutation vector**

$$\Sigma \leftarrow diag\left(\left(\sigma_{on_1}^k\right)^2, \ldots, \left(\sigma_{on_d}^k\right)^2\right) \in \mathbb{R}^{d \times d}$$
$$\rho \leftarrow N(0, \Sigma)$$

**Step 8: Mutation**

$$x_{on}^k \leftarrow x_{on}^k + \rho$$

**Step 9: Evaluation of the offspring in the objective function**

$$fx_{on}^k \leftarrow f(x_{on}^k)$$

**Step 10: Verification of the number of generated particles**

if $n == \lambda$
  Go to step 10
else
  $n \leftarrow n+1$
  Go to Step 4
end

**Step 11: Selection**

$$x_p^k \leftarrow best(x_o^k, \mu)$$

**Step 12: Verify the stop condition**

if $k == k_{max}$
  Best solution $\leftarrow best(x_p^k)$
  End the search process
else
  $k \leftarrow k+1$
  Go to Step 4
end

Additionally, the MATLAB implementation is detailed in Code 10.7. This implementation uses the Rastringin function in two dimensions, with a population size of $\mu = 3$, an offspring size of $\lambda = 6$, 500 iterations, and constant parameters $c_1$ and $c_2$ equal to 2.

```matlab
% Code 10.7
% Adaptive (μ,λ) ES

% Clear memory and close windows
clear all
close all
% Parameter configuration
    mu = 3;             % Number of particles
    lambda = 6;         % Number of children
    d = 2;              % Dimensions
    lb = [-5.2 -5.12];  % Lower bound
    ub = [5.12 5.2];    % Upper bound
    k = 0;              % Iteration counter
    kmax = 500;         % Maximum number of iterations
    c1 = 1;
    c2 = 1;
    tau  = c1*(1/(sqrt(2*sqrt(d))));
    tau_prima = c2*(1/(sqrt(2*d)));
% Optimization problem (minimization), objective function definition
    f = @(x) 10*d + sum(x.^2 - 10*cos(2*pi.*x));
% Initialize particles and standard deviations
    for i = 1:mu
        Xp(i,:) = rand(1,d).*(ub-lb)+lb;
        sigma_p(i,:) = rand(1,d)*2;
    end
% Evaluate initial particles in the objective function
    for i = 1:mu
        fXp(i,1) = f(Xp(i,:));
    end
% Definition of the serach space
    xAxis=linspace(min(lb),max(ub),85);
    yAxis=xAxis;
    zAxis=[];
    for i = 1:length(xAxis)
        for j = 1:length(yAxis)
            zAxis(i,j) = f([xAxis(i) yAxis(j)]);
        end
    end
    [yAxis,xAxis] = meshgrid(xAxis,yAxis);
% Iterative process
while k<kmax
    for n=1:lambda
        % Select parents
            % Parent 1
            a = randi(mu);
            Xpa  = Xp(a,:);
            sigma_pa = sigma_p(a,:);
            % Parent 2
            b = randi(mu);
            Xpb = Xp(b,:);
            sigma_pb = sigma_p(b,:);
        % Recombination
            for i=1:d
                if rand<0.5
                    Xo(n,i) = Xpa(i);
                    sigma_o(n,i) = sigma_pa(i);
                else
                    Xo(n,i) = Xpb(i);
                    sigma_o(n,i) = sigma_pb(i);
```

## 10.5 Variants of Evolutionary Strategies

```
              end
          end
      % Iterative process
          phi = normrnd(0,1);
          phi_prima = normrnd(0,1,1,d);
          sigma_o(n,:) = sigma_o(n,:).*exp(tau*phi +
tau_prima*phi_prima);
      % Generate mutation vector
          rho = normrnd(0,sigma_o(n,:));
      % Mutation
          Xo(n,:) = Xo(n,:) + rho;
      % Verify bounds
          for j=1:d
              if Xo(n,j) < lb(j)
                  Xo(n,j) = rand*(ub(j)-lb(j))+lb(j);
              elseif Xo(n,j) > ub(j)
                  Xo(n,j) = rand*(ub(j)-lb(j))+lb(j);
              end
          end
      % Evaluate the offspring in the objective function
          fXo(n,1) = f(Xo(n,:));
  end
  % Draw the search space
      figure(1);
      surf(xAxis,yAxis,zAxis)
      hold on
  % Draw particles Xp y Xo

plot3(Xo(:,1),Xo(:,2),fXo,'o','MarkerFaceColor','y','MarkerSize',10)

plot3(Xp(:,1),Xp(:,2),fXp,'o','MarkerFaceColor','m','MarkerSize',10)
      pause(0.01)
      hold off
  % Draw contour
      figure(2)
      contour(xAxis,yAxis,zAxis,20)
      hold on
  % Draw Xp y Xo in contour
      plot(Xo(:,1),Xo(:,2),'o','MarkerFaceColor','y');
      plot(Xp(:,1),Xp(:,2),'o','MarkerFaceColor','m');
      pause(0.01)
      hold off
  % Select the best mu individuals
      [fXo, ind] = sort(fXo);
      Xo = Xo(ind,:);
      sigma_o = sigma_o(ind,:);
      Xp = Xo(1:mu,:);
      sigma_p = sigma_o(1:mu,:);
      fXp = fXo(1:mu);
      k=k+1;
end
% Show results
      display(['Best solution: x1= ',num2str(Xp(1,1)),', x2= ',num2str(Xp(1,2))])
      display([' f(X)= ',num2str(fXp(1))])
```

From Algorithms 10.4 and 10.5, which correspond to the $(\mu + \lambda)$ ES and $(\mu, \lambda)$ ES methods respectively, it can be seen that the primary distinction between these and their adaptive versions described in Algorithms 10.6 and 10.7 is the inclusion of the adaptive process. This process involves modifying the standard deviations vector $\sigma_o$ associated with the offspring $x_o$ according to Eq. 10.16 before mutating $x_o$.

To compare the performance of the $(\mu + \lambda)$ ES, $(\mu, \lambda)$ ES, and their adaptive versions, several tests were conducted. Each method was run 100 times using the Rastrigin function in 30 dimensions with a population size of $\mu = 20$, an offspring size of $\lambda = 30$, and 500 iterations. In the adaptive methods, the constant parameters $c_1 = 1$ and $c_2 = 2$ were used. The average results obtained in each iteration by the adaptive $(\mu + \lambda)$ ES and $(\mu + \lambda)$ ES methods are illustrated in Fig. 10.17, while Fig. 10.18 shows the performance of the $(\mu, \lambda)$ ES algorithm and its adaptive version.

It is evident that the performance of the adaptive versions is superior to the original methods, as illustrated in Figs. 10.17 and 10.18.

A comparison between the adaptive $(\mu + \lambda)$ ES and the adaptive $(\mu, \lambda)$ ES methods is shown in Fig. 10.19, clearly indicating that the adaptive $(\mu + \lambda)$ ES algorithm outperforms the adaptive $(\mu, \lambda)$ ES for the Rastrigin function.

In all the variants of Evolutionary Strategy algorithms explored so far, selection plays a crucial role in the success of the search. While the $(\mu, \lambda)$ ES and its adaptive version eliminate each generation of solutions, retaining only the best children, the $(\mu + \lambda)$ ES and its adaptive version retain only the best individuals from the entire population. These mechanisms result in different degrees of exploration and exploitation, which can

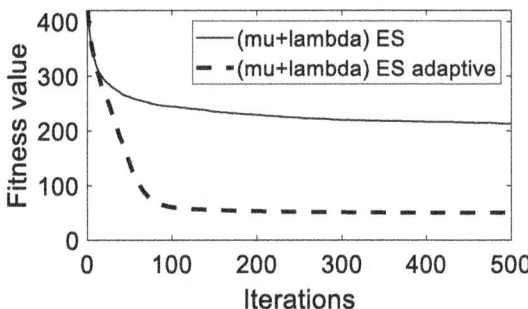

**Fig. 10.17** Performance analysis of the $(\mu + \lambda)$ ES and the adaptive $(\mu + \lambda)$ ES

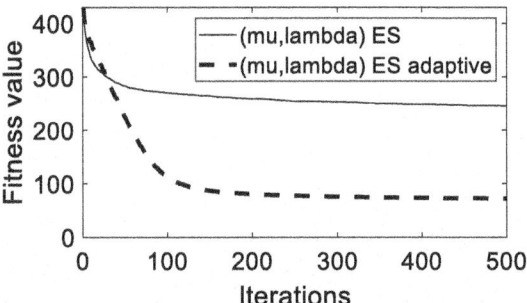

**Fig. 10.18** Performance analysis of the $(\mu, \lambda)$ ES and the adaptive $(\mu, \lambda)$ ES